Therapeutic Applications of Quadruplex Nucleic Acids

T0311892

Therapeutic Applications of Quadruplex Nucleic Acids

Stephen Neidle

The School of Pharmacy,
University of London

AMSTERDAM • BOSTON • HEIDELBERG • LONDON
NEW YORK • OXFORD • PARIS • SAN DIEGO
SAN FRANCISCO • SINGAPORE • SYDNEY • TOKYO
Academic Press is an imprint of Elsevier

Academic Press is an imprint of Elsevier
32 Jamestown Road, London NW1 7BY, UK
225 Wyman Street, Waltham, MA 02451, USA
525 B Street, Suite 1800, San Diego, CA 92101-4495, USA

First edition 2012

Notice
No responsibility is assumed by the publisher for any injury and/or damage to persons or property as a matter of products liability, negligence or otherwise, or from any use or operation of any methods, products, instructions or ideas contained in the material herein. Because of rapid advances in the medical sciences, in particular, independent verification of diagnoses and drug dosages should be made

British Library Cataloguing-in-Publication Data
A catalogue record for this book is available from the British Library

Library of Congress Cataloging-in-Publication Data
A catalog record for this book is available from the Library of Congress

ISBN : 978-0-12-810372-2

For information on all Academic Press publications
visit our website at elsevierdirect.com

Typeset by MPS Limited, a Macmillan Company, Chennai, India
www.macmillansolutions.com

Printed and bound in United States of America

10 11 12 13 14 15   10 9 8 7 6 5 4 3 2 1

Working together to grow
libraries in developing countries

www.elsevier.com | www.bookaid.org | www.sabre.org

ELSEVIER   BOOK AID
International   Sabre Foundation

Dedication

To
Dan and Leora, Ben and Natalie, Hannah and Mark

Contents

Preface

For those of us studying nucleic acids, a sense of history comes with the territory. My own induction in the field started in the 1970s at King's College London, where I soon learnt that the determination and analysis of nucleic acid structure continued to have its controversies and rivalries, not least in the area of tRNA that I was involved in for a short time as a very junior player. I also heard from many quarters that DNA structure was no longer interesting, unlike proteins, although RNA was different. It took some while for this prevailing view to change, that there was little new to be learned about DNA structure after the double helix. The seminal studies by David Davies and colleagues in 1968, which showed for the first time that a non-duplex four-stranded type of structure could be formed by guanine-containing nucleic acid components, were largely ignored for almost 20 years. The realization that guanine-rich sequences of DNA could fold up into stable structures, rather like a polypeptide, only became of wider interest when a biological role for these structures in eukaryotic telomeres was proposed in 1988–89, and NMR/crystal structures started to emerge for these novel quadruplex arrangements.

The quadruplex field received a further impetus a few years later; first, arising from the seminal discovery of the telomerase enzyme complex by Elizabeth Blackburn and colleagues; and later from the demonstration by Shay, Weinberg and many others that telomerase plays a major role in human cancer since it is the driver for telomere maintenance in most cancer cells. I was fortunate to be involved in some of the early quadruplex studies that grew from the telomerase area, when in collaboration with my good friend and colleague Laurence Hurley, the proof of principle was demonstrated that small-molecule mediated induction of quadruplex formation is an effective way of inhibiting telomerase function. Since then a very large number of laboratories have extended these findings, chemically, biologically and structurally. Many are striving to develop this approach into an effective anti-cancer strategy. A notable success was achieved with Quarfloxin™, the first quadruplex-targeted drug molecule to be entered into a clinical trial for cancer. Others will undoubtedly follow, even in the current challenging economic climate. There continues to be interest in quadruplexes in both large pharma and in small drug discovery companies, although most of the activity in quadruplex therapeutics to date has come from academia. The model for successful drug discovery is rapidly changing and quadruplex therapeutics is an ideal exemplar for the future, hopefully with the successful translation of academic discoveries into industry and the clinic.

This book is intended for all who are interested in quadruplex nucleic acids as new targets in disease treatment and drug discovery. I hope that it will appeal to those who are actively involved in the area. The book covers the underlying chemistry and biology especially as applied to small-molecule drug discovery, but emphasizes throughout my own interests and perspective in molecular structural results, concepts and principles as applied to quadruplex-targeted drug discovery. I have not attempted to comprehensively review all of the literature; this is no longer possible within the constraints of a small book in view of the continued rapid growth of the subject and the large number of existing publications in the quadruplex field—over 2400 by May 2011. I have deliberately referenced a number of the many excellent reviews on the chemistry and biology of quadruplexes, especially those that provide greater detail in individual topics than this book can include.

I am grateful to many friends and colleagues worldwide for their collaborations, discussions and insights over the past 14 years. Special mention goes to Laurence Hurley (Arizona), Shankar Balasubramanian (Cambridge, UK), Dinshaw Patel (New York), Struther Arnott, the members of my own group in London who continue to generate excellent science and a stimulating environment, and my colleagues at the School of Pharmacy for their continuing indulgence. Cancer Research UK has been a generous funder of much of our work over many years, and I am profoundly grateful to it and its countless supporters. My wife Andrea, as ever, has been a constant source of inspiration and encouragement throughout this project.

Diana V. Silva and Jose M. Rivera (University of Puerto Rico) are thanked for providing this version of a cartoon that has appeared on their excellent blog http://web.mac.com/jmrivortz/JMR-Lab/Blog/. Any resemblance to the author is entirely intentional!

Stephen Neidle

1

Introduction: Quadruplexes and their Biology

This book is organized so that the reader can progress through accounts of the fundamentals of quadruplex biology and three-dimensional structure, through to the chemistry and biology of quadruplex interactions with small molecules. The functionally and structurally distinct telomeric, genomic and RNA quadruplexes are each discussed in separate chapters. The book continues with an extended discussion of the major issues and challenges for quadruplexes as therapeutic targets. The underlying theme throughout the book is molecular structure and in particular how structural concepts can be applied to quadruplex-targeted ligand design and discovery. The concluding chapter provides background material for this, including the basics of diffraction and crystallographic techniques as applied to quadruplex nucleic acids.

Helical Arrangements of Guanosine Repeats

The observations that guanine-rich nucleic acids, and indeed guanosine itself, have unusual physical properties, are exactly 100 years old. It was first noted by Bang (1910) that these readily formed gel-like substances in aqueous solution. Analogous findings of (highly ordered) aggregation were made with the first oligonucleotides containing deoxyguanosine to have been synthesized (Ralph, Connors & Khorana, 1962). These workers also reported some of the first spectroscopic studies on these unusual nucleic acids and showed that the optical density of the tri- and tetradeoxynucleotides d(pGGG) and d(pGGGG) changed with temperature to give a sharp thermal transition point (T_m), indicative of an ordered secondary structure, although the authors did not speculate on its nature. These observations were rationalized in structural terms by the systematic X-ray fiber diffraction study of Gellert, Lipsett and Davies (1962), who studied fibers formed from gels of 3′- and 5′-guanosine monophosphate, analogous to the original gels produced some 50 years previously by Bang. The diffraction patterns showed a number of the characteristics of helical nucleic acid structural arrangements seen a decade earlier with diffraction patterns of fibrous random-sequence double-helical DNA by Franklin and Wilkins, with, in particular, a strong meridional reflection at 3.3Å indicative of stacked bases, although the dimensions of the GMP quasi-helices are distinct from those of the double helix. Gellert *et al.* also pointed out that the regular structure and high stability apparent in these helices could be explained by a hydrogen-bonding arrangement of four guanine bases (subsequently termed the G-quartet or G-tetrad), with two

Therapeutic Applications of Quadruplex Nucleic Acids. DOI: 10.1016/B978-0-12-375138-6.00001-7

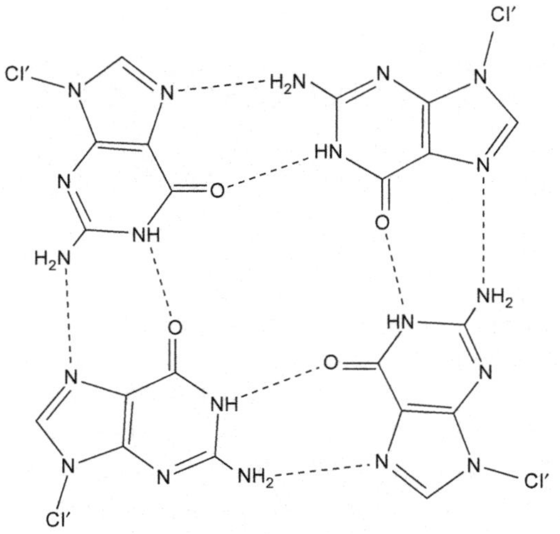

FIGURE 1–1 The structure of the G-quartet, showing the hydrogen bonding arrangement between the four coplanar guanine bases.

hydrogen bonds between each pair (Figure 1–1) involving four donor/acceptor atoms of each guanine base: the N1, N7, O6 and N2 atoms (Davis, 2004). The four-fold symmetrical G-quartet arrangement was based on a dimerization of a guanine–guanine base pairing earlier suggested by Donohue (1956). Subsequent fiber-diffraction studies, discussed in Chapter 2, have confirmed and extended this model.

The Rise of the Quadruplex Concept

These early observations and rationalizations of guanine aggregation were given new impetus by subsequent findings over two decades after the initial fiber-diffraction study that guanine-rich sequences in immunoglobin switch regions (Sen & Gilbert, 1988) and in telomeric regions at the ends of eukaryotic chromosomes (Henderson *et al.*, 1987; Sundquist & Klug, 1989) can also form this type of four-stranded structural arrangement. Such sequences have discrete runs of guanine tracts, which do not form continuous helices but instead comprise compact structured arrangements, which it was soon realized can also be formed by short-length oligonucleotides of appropriate sequence. These are termed **quadruplexes** (the name tetraplex is occasionally also used), and can be constructed from one, two or four strands of (ribo- or deoxy-) oligonucleotide (Figure 1–2). Quadruplexes are thus structures containing as their basis typically 3–4 G-quartets linked together, and can be either discrete entities formed by short sequences or embedded within a longer nucleic acid sequence.

This chapter summarizes the underlying biology of telomeres and how telomeric quadruplexes may be involved in telomeric processes, as well as other telomere-related therapeutic directions. Guanine-rich sequences in DNA have also been identified in a number of other genetic contexts notably in oncogene promoter regions (Murchie

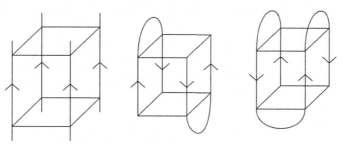

FIGURE 1–2 Schematic showing three different types of quadruplex arrangement, with four, two and one oligonucleotide strand, respectively. Strand directions are shown by arrows.

& Lilley, 1992; Simonsson, Pecinka & Kubista, 1998; Simonsson, 2001). More recently many other sequences have been identified as putative quadruplex forming at the DNA and increasingly at the RNA level, using bioinformatics approaches and knowledge of the annotated human genome sequence to locate them. In turn the potential of these quadruplex nucleic acid structures for targeted gene regulation at the transcriptional and translational levels, both endogenous and ligand-aided, has generated much interest and activity. These topics are described in subsequent chapters of this book.

Guanine-rich Sequences in Telomeres

Eukaryotic organisms have evolved a common mechanism to protect the ends of their chromosomes from unwanted recombination, fusions, or nuclease attack (Blackburn, 1991; Cech, 2000, 2004), by forming specialized structures at chromosome ends. These structures, known as telomeres, consist of repetitive, guanine-rich (telomeric) DNA together with an array of telomeric proteins. The length and sequence of telomeric DNA depends on the nature of the organism. Yeast organisms have complex sequence repeats of the type $(TG)_{1-6}TG_{2-3}$ (*S. cerevisiae*) and $TTACAG_{1-8}$ (*S. pombe*), and the well-studied ciliate *Oxytricha nova* has the repeat TTTTGGGG. Vertebrate telomeres contain the hexanucleotide repeat 5′-TTAGGG (Moyzis *et al.*, 1988; Meyne *et al.*, 1989), although there are wide species-dependent variations in the length of their telomeric DNA. Human telomeric DNA sequences typically range in length from 3–4 up to ca 15 kilobases, contrasting with mouse telomeres, which average 50 kilobases in length. All except the terminal 3′ end 15–200 nucleotides are in a DNA duplex form, with complementary sequence repeats 3′-AATCCC. The 3′ extreme end of a chromosome is uniquely single-stranded, and is often termed the single-stranded overhang (Wright *et al.*, 1997). It is capable in principle of folding into quadruplex structures, although it is normally associated with proteins, notably several copies of the single-stranded telomeric DNA-binding protein hPOT1 (Baumann & Cech, 2001; Lei, Podell & Cech, 2004).

Human telomeric DNA does not form a conventional chromatin-type structure, as is the case for most genomic DNA. We do not have structural data comparable to that obtained for the nucleosome (Schalch *et al.*, 2005: PDB id 1ZBB), although electron microscopy data are available which suggests that some telomeres end in a single large loop structure (Griffith *et al.*, 1999), at least for part of the cell cycle. The telomere loop structure comprises a large t-loop, visible in electron micrographs, which comprises duplex telomeric DNA, together with a small region, termed the D-loop, where the single-strand overhang DNA end is embedded into the t-loop (Figure 1–3). The detailed arrangement of the D-loop is unknown and several theoretical models have been devised, some of which include a four-way junction or a quadruplex arrangement at the point where the overhang is presumed to interact with the duplex. The loops are associated with a large number of proteins, most of which are specific for the telomere (Figure 1.3). A number of copies of the TRF1 and TRF2 (telomere repeat binding factor) proteins bind to duplex telomeric DNA and are linked together by the TIN2 protein. TRF2 is linked to the single-stranded overhang region and POT1 by the TPP1 protein. The complete assembly is termed the shelterin complex (de Lange, 2005, 2010), and its role is to regulate telomeric DNA responses to potentially lethal damage events, as well as playing a role in telomeric DNA length regulation.

Telomeric DNA naturally shortens during each round of cellular replication as a consequence of the inability of the DNA polymerase replication machinery of the cell to fully replicate the blunt ends—the so-called "end-replication" problem (Cech, 2004). Thus in the absence of any compensating mechanism, telomeres in normal cells progressively

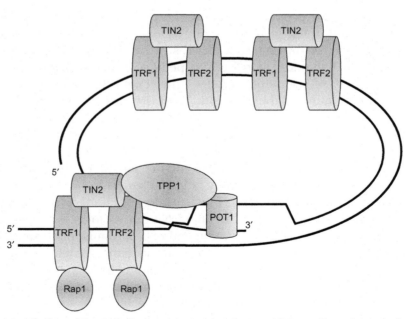

FIGURE 1–3 Schematic view of the shelterin complex, showing the t- and D-loops, the various associated proteins and their mutual interactions.

shorten until they reach a critical point, the "Hayflick limit," when cells respond by ceasing replication and enter the senescent state of replication halt, which is normally irreversible, and which then can lead to cellular apoptosis (Zhang *et al.*, 1999; Shay & Wright, 2010a, b). Short telomeres are also signals for invoking a DNA damage response, which similarly leads to senescence and cell death (d'Adda di Fagugna *et al.*, 2003; d'Adda di Fagagna, Teo & Jackson, 2004; Reaper *et al.*, 2004). Exposure of the 3′ ends of telomeric DNA, for example by removing the POT1 protein, also leads to the induction of a senescence pathway (Li *et al.*, 2003).

Given that the telomeric ends of chromosomes are single stranded, and that G-tract sequences have an innate propensity to fold into quadruplex structures, the question has been asked as to whether telomeric quadruplexes occur naturally, whether they have particular cellular functions, and thus whether they are stable in a cellular environment (Maizels, 2006: Oganesain & Bryan, 2007). Direct evidence for the existence of parallel telomeric G-quadruplexes has been obtained (Schaffitzel *et al.*, 2001) from a ciliate (the organism *Stylonychia lemnae*) by antibody staining. More direct functional evidence for the existence of these structures has also come from studies on ciliates (Paeschke *et al.*, 2008; Lipps & Rhodes, 2009), which have a high proportion of their genome consisting of telomeric DNA, making experiments much more straightforward to undertake and interpret than in mammalian cells. These studies have shown that the telomere-binding proteins TEBPα and TEBPβ cooperate in binding and stabilizing G-quadruplexes, which would spontaneously form in ciliates, but not in mammalian cells due to the presence of the POT1 single-strand binding protein. More recently, direct evidence of the existence of G-quadruplexes in mammalian cells has emerged from a pull-down assay approach (Müller *et al.*, 2011) which has used a small-molecule ligand with high quadruplex-binding affinity and specificity—it has low affinity for duplex DNA. The ligand, attached to beads, was incubated with genomic DNA fragments isolated from a human cancer cell line, and it was found that the DNA pulled down by the beads was exclusively telomeric DNA. This result is consistent with the ligand specifically targeting telomeric quadruplex structures and not single-stranded or duplex telomeric DNA. Quadruplex structures also bind and catalytically activate (Soldatenkov *et al.*, 2008) the human nuclear protein poly (ADP-ribose) (PARP), which is involved in DNA repair. The further consequences of this are discussed in Chapter 6.

Telomere Maintenance and Telomerase

The seminal finding of an enzymatic activity in the ciliate *Tetrahymena* that can elongate telomeric DNA and maintain telomere homeostatis, by adding telomeric repeats, was made by Greider and Blackburn (1985). An enzyme complex with this activity was isolated and termed telomerase (Mergny *et al.*, 2002; Autexier & Lue, 2006: Blackburn & Collins, 2011; Wyatt, West & Beattie, 2010). This enzyme reverses the behavior of telomeric DNA in somatic cells, which progressively shorten during replication, enabling telomerase-positive cells to bypass replicative senescence and to achieve cellular immortality. Telomerase is expressed in the nucleus of many eukaryotic organisms, albeit in almost all instances in very low copy numbers. It consists of two principal subunits, a catalytic domain (TERT: hTERT in humans: Meyerson *et al.*, 1997; Nakamura *et al.*, 1997) and an RNA domain (TR: hTR in humans: Feng *et al.*, 1995; Greider & Villeponteau, 1995) containing a single-stranded RNA template which serves to recognize the 3' end of telomeric DNA. The telomerase catalytic subunit, which has reverse transcriptase activity and sequence homology to viral reverse transcriptases, is responsible for the addition of deoxynucleotide triphosphates to the 3' end of the telomeric DNA substrate at the catalytic site, which thus has to be positioned close to the RNA template for efficient nucleotide addition to occur.

A crystal structure of the full-length telomerase catalytic domain TERT from the beetle *Tribolium castaneum* has been determined (Gillis, Schuller & Skordalakes, 2008; PDB ids 3DU5, 3DU6); although no equivalent structure is available for the human enzyme, this TERT has high homology with it. TERT is organized with four domains into a ring-like structure, one of which, the palm domain, contains the catalytic site. More recently the crystal structure of the same TERT bound to a short RNA-DNA hairpin (Mitchell *et al.*, 2010; Mason, Schuller & Skordalakes, 2010; PDB id 3KYL) has been reported (Figure 1–4a, b), which shows that nucleic acid binding induces a series of changes in the relative positions of the domains. Although this model sequence can only approximate the natural human telomerase RNA hTR, which has 451 nucleotides, its complex with TERT shows that the mechanism of nucleotide addition to telomere ends is likely to be closely analogous to that of viral reverse transcriptases, although a major difference with these enzymes is that telomerase uses an endogenous 11-nucleotide RNA template of complementary sequence r(AAUCCCAAUC) (Feng *et al.*, 1995) on which the DNA telomeric repeat TTAGGG is synthesized.

Human telomerase expression was first identified in cancer cells (Counter *et al.*, 1994; Kim *et al.*, 1994), following on from the findings that (i) telomeric DNA in many cancers is shortened yet stable in length relative to somatic cells and (ii) that telomerase is not normally expressed in human somatic cells (though it may be expressed at low levels in human embryonic and stem cells). The intimate association between cancer and telomerase was demonstrated by a large number of subsequent studies on cancer cell lines and material from primary tumors in patients (Shay & Bacchetti, 1997), concluding that overall 80–85% of human cancers express significant levels of enzymatically active telomerase. Telomerase plays a key role in cellular immortalization and has been identified as one of the essential factors required in order to transform a normal cell to an immortalized one (Bodnar *et al.*, 1998; Hahn *et al.*, 1999a; Hanahan & Weinberg, 2000, 2011). Few cancer types fail to express telomerase. These total some 10–15% of mostly rarer cancers, notably soft tissue sarcomas and osteosarcomas. They maintain telomere length by ALT (Alternate Lengthening of Telomeres) pathways, which appear to involve homologous recombination rather than enzymatic generation of telomeric DNA (Cesare & Reddel, 2010).

Inhibition of telomerase catalytic activity leads to a progressive loss of TTAGGG repeats at the 3' ends of telomeres in human cells and thus to progressive telomere

(a)

(b)

FIGURE 1–4 (a), (b) Two views of the crystal structure of the TERT telomerase sub-unit from *T. castaneum* with a bound DNA-RNA hairpin (Mitchell *et al.*, 2010).

shortening since as a consequence of the end-replication effect this loss is not replaced during replication. Telomeres in telomerase-negative cells become gradually shortened, and once telomeric DNA is at a critically short length, cells then enter Rb-dependent replicative senescence (the irreversible loss of cell division), and ultimately apoptosis. Telomere shortening in the absence of significant telomerase expression appears to be a tumor suppressor mechanism. The cellular consequence of shortening is in principle a lengthy process, dependent on initial telomere length and replication time for a particular cell type. Thus the classic time-lag model of telomerase inhibition and consequent telomere attrition requires that cells with a mean telomere length of 5 kb, a 24 h cell doubling time and a subsequent loss of ~100 nucleotides per round of replication would reach critical telomere shortening in ~40–50 days (Figure 1–5: Kelland, 2007; Oganesian & Bryan, 2007). A key set of findings has been that inhibition of hTERT, for example by siRNA, antisense or small-molecule inhibitors, not only reverses telomere maintenance, but also selectively inhibits cancer cell growth (Hahn *et al.*, 1999b; Herbert *et al.*, 1999; Zhang *et al.*, 1999; Kosciolek *et al.*, 2003) and strongly supports the broad concept that induction of telomere shortening is a viable therapeutic strategy. It was initially envisaged that the large time lag for the onset of senescence and apoptosis would present a practical challenge to the eventual clinical use of telomerase inhibitors, a view initially supported by data on direct telomerase inhibitors. As we will see below and in Chapter 6, the actual biology of telomerase inhibitors, and in particular quadruplex inhibitors, has been found to follow a distinct and potentially more clinically useful set of pathways.

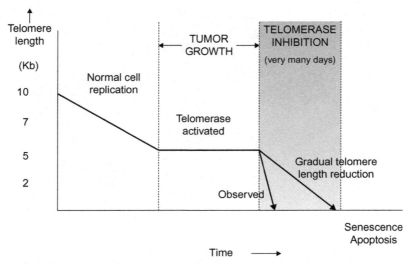

FIGURE 1–5 Schematic representation of the pathway of events that occur during the lifespan of a human somatic cell, going from progressive shortening through to a transformation event, when telomere stabilization occurs and telomerase is activated, through to telomerase inhibition. The last set of events show both the classic model of gradual telomere length reduction and the actual observations with G-quadruplex inhibitors, of a more rapid response.

Telomerase Targeting for Therapy

The concept of telomerase as a therapeutic target in human cancer has resulted in several distinct approaches to telomerase inhibition. Historically the first was the screening of nucleotide-based reverse transcriptase inhibitors of hTERT catalytic function (Strahl & Blackburn, 1994, 1996), including the HIV drug AZT (3′-azido-2′,3′-dideoxythymidine; Melana, Holland & Pogo, 1998). None of these compounds have advanced to clinical evaluation probably because of problems of potency, selectivity and toxicity. A number of non-nucleoside synthetic inhibitors, typified by the compound BIBR1532, have received more recent attention (Damm *et al.*, 2001; Pascolo *et al.*, 2002). BIBR1532 is a potent non-competitive telomerase inhibitor *in vitro*, with a $^{tel}IC_{50}$ value of 93 nM (Damm *et al.*, 2001), although its activity in cells is 50-fold lower (Barma *et al.*, 2003). However, the behavior of this compound in experimental cancer models has not encouraged further activity in this compound type. Critical telomere shortening was only observed after 120 days of exposure in the NCI-H460 non-small cell lung cancer cell line (Damm *et al.*, 2001). This behavior of BIBR1532 corresponds to a classic model of telomerase inhibition, with an extended time lag before telomeres reach critical shortening. Also in accordance with this model, *in vivo* anti-tumor activity was only observed with an animal xenograft tumor model after >100 days of pre-treatment of the cancer cells to shrink telomere length prior to establishing the xenograft. Further development of BIBR1532 was discontinued since the extended time-lag behavior is not therapeutically viable in humans, although it was subsequently found that much higher doses were selectively cytotoxic in leukaemia cells as a result of telomere damage rather than a mechanism involving inhibition of catalytic activity (El-Daly *et al.*, 2005).

Inhibition of the template function of the RNA component (hTR) of telomerase can be achieved by oligonucleotides with sequence complementary to the template, which has resulted in a more fruitful direction for eventual therapeutic development. This inhibition can result in potent inhibition of hTERT catalytic activity since template recognition by the 3′ end of the telomeric DNA overhang is the essential first step in the elongation cycle. One such oligonucleotide has progressed to clinical trial stages in cancer and is currently in Phase II evaluation (see www.geron.com), and its potential for targeting cancer stem cells (which express telomerase) is also being studied (Joseph *et al.*, 2010; Marian *et al.*, 2010). This 13-mer oligonucleotide, termed GRN163L (Imetelstat), has the backbone modified by thiophosphoramidate groups to enhance cellular stability and complementary RNA affinity together with a 5′ palmitoylated tail to enhance uptake. The GRN163L sequence d(TAGGGTTAGACAA) binds with high affinity to the RNA template but cannot by itself form stable quadruplex-like secondary structure. The drug rapidly reduces telomerase activity in target cells, within 3–4 days, but telomere shortening is reduced only after the calculated time lag. It does have significant *in vivo* anti-tumor activity in a wide range of solid tumor and hematological cancer models, which is telomere shortening independent (Röth, Harley & Baerlocher, 2010). An important recent finding (Castelo-Branco *et al.*, 2011) has been that Imetelstat is able to effectively target neural tumor-initiating (i.e. tumor stem cells) in a xenograft model while at the same time normal tissue stem cells were unaffected. The potential for toxicity to normal stem cells has been a concern for all categories of telomerase inhibitors, and this study suggests that such concerns are unfounded.

A third approach to telomerase inhibition, and the one that is the focus of much of this book, is more indirect. It is based on the seminal finding (Zahler *et al.*, 1991) that folding of the telomeric DNA primer into a quadruplex structure inhibits telomerase activity (Wang *et al.*, 2011). However, this does not occur naturally since in cells the single-stranded telomeric DNA overhang is associated with protective proteins, notably POT1 (Zaug, Podell, & Cech, 2005). It was realized, prior to the discovery of these proteins, that this process would require a driving force to stabilize quadruplex folding. It was proposed that a suitable quadruplex-binding small molecule could do this and thus act as an indirect telomerase inhibitor (Figure 1–6). The proof of principle experiments that demonstrated the feasibility of the approach used a quadruplex-binding bis-amido-substituted anthraquinone derivative (Sun *et al.*, 1997). Telomerase inhibitory potency was determined by a direct assay to be $23\,\mu M$ ($^{tel}IC_{50}$ value) (this value cannot be compared with those from more recently discovered ligands in view of the differences, not only in assay type, but in the sequence of primer used). Many laboratories have built on and extended this study, with the subsequent

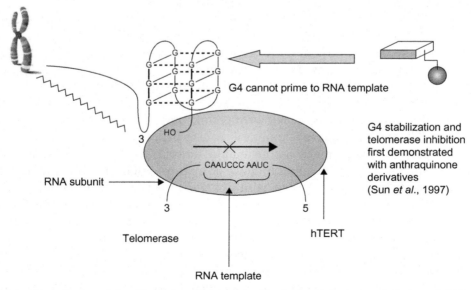

The 3′ end of a telomeric DNA primer strand is required to be single-stranded in order to effectively hybridize with the 11 bp template of hTR.

Inducing it to fold into a quadruplex structure will inhibit hybridization, and thus half further telomere elongation and protein binding

G4 cannot prime to RNA template

G4 stabilization and telomerase inhibition first demonstrated with anthraquinone derivatives (Sun *et al.*, 1997)

RNA subunit

CAAUCCC AAUC

Telomerase

hTERT

RNA template

FIGURE 1–6 Schematic view of the effects of quadruplex formation in the 3′ single-stranded telomeric DNA overhang on telomerase function. The quadruplex structure requires stabilization by a small-molecule ligand.

discovery of a large number of quadruplex-binding telomerase inhibitors (Chapters 4 and 5). Telomere shortening has been observed for a number of compound types, but many also produce profound short-term cell growth inhibition that cannot be due solely to the classic telomerase model for the action of these molecules (Figure 1–5), which is fortunate, since otherwise their clinical potential as a class would be very restricted at best. One plausible explanation for these findings is that within a population of cells there will be some with exceptionally short telomeres—these are especially sensitive (Hemann *et al.*, 2001; Ijpma & Greider, 2003; Feldser & Greider, 2007) and the survival of the overall mass of cells may well be dependent on this subpopulation. This subpopulation of cells will require very few rounds of replication following treatment with a telomerase inhibitor for their senescence to be activated, with potentially profound effects on the whole cell population. Other non-telomerase responses to these telomere targeting compounds are discussed in Chapter 6.

Telomerase Inhibition Measurement

The enzymatic activity arising from the catalytic function of telomerase hTERT is in principle straightforward to assay. All variants of the assay add telomeric TTAGGG repeats to the 3′ end of a DNA primer (which can be the telomeric DNA substrate) using nucleotide triphosphates. The readout is the sequence of the elongated primer and this can be estimated in several ways—often it is visualized on a non-denaturing polyacrylamide gel as a ladder of bands (Figure 1–7), each of which corresponds to an added TTAGGG repeat. A reduction in bands compared to a control indicates enzyme inhibition. Early "direct" assays (see, for example, Sun *et al.*, 1997) used an isotopic approach with [α-^{32}P] dGTP nucleotide triphosphate having high specific radioactivity, but this methodology is not now used because of the unacceptably high level of radiation needed. Improvements

FIGURE 1–7 A non-denaturing polyacrylamide gel from a TRAP-LIG experiment examining the effects of different ligand concentration on the activity of telomerase. The lanes with differing concentrations of ligand (AZT) are marked, and the pattern of bands is seen to become progressively weaker and less extended with increasing concentration. Each band corresponds to a TTAGGG addition on to the telomeric DNA primer. I am grateful to Dr. M. Gunaratnam for supplying this picture.

+   15   25   30   50   neg

have been made on the direct assay, for example, using biotinylated primer so that products can be isolated on streptavidin-coated beads (Sun, Hurley & von Hoff, 1998), which could in principle be used in a high-throughput assay.

The reliable measurement of telomerase inhibition, especially by small molecules, has thus been a technical challenge, especially since the very low copy number of telomerase in human cancer cells has meant that a direct assay cannot be used without an amplification step. Normal enrichment of extracts from a telomerase-positive cancer cell line such as the human breast carcinoma line MCF7 does not produce sufficient enzyme for the direct assay without the need to have $[\alpha\text{-}^{32}P]$ dGTP in the assay. Instead a PCR-amplification procedure has been developed (Wright, Shay & Piatyszek, 1995; Kim & Wu, 1997), the TRAP (Telomerase Repeat Amplification Protocol), which is now almost universally used for analyzing telomerase activity *in vitro*, in cells and in tissues. Several variants of the TRAP assay are available commercially (for example, the TRAPEZE™ kit from www.millipore.com). The TRAP assay uses two steps; first, addition of telomere repeats to a primer sequence by telomerase, and second, amplification of the products by repeated PCR cycles. This is followed by readout of the PCR products, commonly accompanied by visualization and quantitation by staining and imaging of the gel. A desirable goal with a small-molecule inhibitor is a dose–response curve from which the concentration required for 50% telomerase enzyme inhibition can be obtained—the $^{tel}IC_{50}$ value. One modification of the assay (Gomez, Mergny & Riou, 2002) has been to use a substrate that can form an intramolecular quadruplex, enabling discrimination between quadruplex-based and enzymatic telomerase inhibitors.

In general the assay is prone to a number of problems such as artifacts of amplification and unreliability of the PCR step, with ligand either directly inhibiting the *TAQ* polymerase enzyme used in the PCR step, or stabilizing secondary structure. Thus reproducibility and reliable quantitation of inhibition data have been challenging issues to overcome. An important critical examination of the assay (De Cian *et al.*, 2007) has concluded that the TRAP assay, even though it has been very widely used in studies of G-quadruplex ligands, is fundamentally flawed since the PCR 2nd step is inherently prone to G-quadruplex ligand carryover and interference. These authors developed a direct assay using cell extracts with high telomerase activity and transient over-expression of hTERT and hTR, thus avoiding the need to use high levels of radioactive ^{32}P dGTP. It was also pointed out that the $^{tel}IC_{50}$ values obtained from this assay are still dependent on primer sequence, which has to be borne in mind when comparing results from different laboratories. Over-expressing hTERT/hTR cell lines are not generally available so a modification of the TRAP procedure has been developed, the TRAP-LIG method, which adds an additional step to the procedure, prior to the PCR amplification, enabling ligand to be fully removed using a commercially available oligo-nucleotide purification kit (Reed *et al.*, 2008). The results are reproducible for a given primer sequence, and agree with the conclusion from the direct assay (De Cian *et al.*, 2007) that the conventional TRAP assay can considerably overestimate $^{tel}IC_{50}$ values for small-molecule G-quadruplex ligands.

Quadruplex Sequences in Genomes

So far we have focused exclusively on telomeric DNA and the consequences of it forming, or being induced to form, quadruplex structures. However, these structures are not restricted to telomeres. The existence of G-tract sequences within a variety of non-telomeric genes was recognised 20 years ago (Simonsson, 1991). The availability of the three billion nucleotide human genome sequence in 2001, followed by that of many other organisms, has provided the raw data for which the existence and identity of such sequences can be systematically searched (D'Antonio & Bagga, 2004; Todd, 2007; Ryvkin *et al.*, 2010). Two separate bioinformatics analyses employing distinct algorithms (Huppert & Balasubramanian, 2005; Todd, Johnson & Neidle, 2005; Todd, 2007) have used the ENSEMBL genome browser as a starting point to examine the prevalence of potential quadruplex sequences in the human genome. The generalized quadruplex probe sequence used was:

$$G_m X_n G_m X_o G_m X_p G_m$$

where m is the number of G residues in each short G-tract and X_n, X_o and X_p can be any combination of residues, including G. This probe sequence does not assume that all the G-tracts are of equal length since guanines can be immediately adjacent to a G-tract. Limits were set for the length of the G-tracts and the intervening sequences (loops):

$$3 \leq m \leq 5$$
$$1 \leq X \leq 7$$

These analyses arrived at remarkably similar conclusions, with totals of ~376,000 (Huppert & Balasubramanian, 2005), and ~375,000 (Todd, Johnson & Neidle, 2005), indicating that there are highly significant and non-random enrichments of quadruplex sequences. Of these, the largest category is in intergenic regions (~223,000), with ~151,000 within genes and just ~14,000 within exons. The question of which of these sequences are capable of forming quadruplexes cannot be directly answered at present; as we shall see the available experimental data on quadruplex structure and stability is minute by comparison. One way to approach the problem is to search for similarities between sequences, to ascertain whether there are families of sequence. Analysis of both overall and loop sequence shows that there are clusters of similar sequence that may well have similar tertiary folds (Todd, Johnson & Neidle, 2005; Todd & Neidle, 2011). Whether this generalized probe sequence encapsulates all possible quadruplex types is open to question.

A number of informatics resources are now available on the web, providing online search facilities and databases of quadruplex motifs. These are summarized in Table 1–1. These generally also use the probe sequence described above, together with its limits on G-tract and loop sizes. Studies on non-mammalian organisms have similarly found enrichment of quadruplex motifs, notably in the *E. coli* genome (Rawal *et al.*,

Table 1–1   Some G-quadruplex search programs and databases

Program Name	Web Address	Download or Server?	Target Genome	Reference
QuadDB	http://www.quadruplex. org/?view=quadparser	both	any	Huppert & Balasubramanian, 2005
QGRS Mapper	http://bioinformatics. ramapo.edu/QGRS/ index.php	server	any	Kikin, D'Antonio & Bagga, 2006
GRSDBGRS_UTRdb	http://bioinformatics. ramapo.edu/GRSDB2/	server	any	Kikin et al., 2008
Greglist	http://tubic.tju.edu.cn/ greglist/	server	several	Zhang, Lin & Zhang, 2007
QuadBase	http://quadbase.igib. res.in/	server	many	Yadav et al., 2008

2006) and in the yeast *Saccharomyces cerevisiae* (Hershman *et al.*, 2007). More detailed analyses in regulatory regions (discussed further in Chapter 7) have shown that there are significantly higher frequencies of such sequences in promoter regions (Huppert & Balasubramanian, 2007) and in introns, with the frequency increasing in tandem with closeness to transcription start sites (see, for example, Eddy & Maizels, 2006, 2008, 2009). The functional consequences of these non-random occurrences are not as yet clear, but a study in *Saccharomyces cerevisiae* across seven strains of this organism has shown that the quadruplex motifs in them are mostly evolutionarily conserved, with high statistical significance (Capra *et al.*, 2010), a good indication of their functional significance.

A strong association was also found in the *Saccharomyces cerevisiae* study for quadruplex occurrence with sites of double-strand DNA breaks. This complements experimental findings in mammalian systems that the FANCJ and other helicases (such as Bloom's syndrome) can unwind quadruplex structures (Sun *et al.*, 1998; Sun, Yabuki & Maizels, 2002; London *et al.*, 2008; Wu *et al.*, 2008; Johnson *et al.*, 2010); these enzymes are involved in the repair of otherwise lethal double-strand DNA lesions. It has been suggested that the function of these particular targeted quadruplex sequences is to regulate gene expression, although it is also clear that their occurrence in particular genes can also be an otherwise lethal event that has to be resolved by, for example, a helicase. The unwinding of the Bloom's and Werner's syndrome helicases can be inhibited by quadruplex-binding di- and trisubstituted acridine derivatives (Li *et al.*, 2001), which may have relevance to the activity of these compounds in telomerase-independent ALT-pathway recombination maintenance of telomere length, which require helicase activity to resolve recombination intermediates.

Other instances of quadruplex sequence association with particular genes are increasingly being reported. Mutations in the ATR-X syndrome protein cause mental retardation and down-regulation of α-globin expression. The protein has been found to preferentially bind to both non-telomeric tandem G-rich DNA repeats (as well as telomeric repeats), and these can form quadruplex structures (Law *et al.*, 2010), suggesting that a major role of this protein is to recognize non-duplex DNA. Another recent example comes from the pathogen field, where it has been found that the recombination sequence used by the pathogen *Neisseria gonorrhoeae* to enhance antigenic variation contains a quadruplex sequence (Cahoon & Seifert, 2009). Biophysical methods have shown that this sequence forms a stable quadruplex and incubation with a quadruplex-binding porphyrin derivative at a non-toxic concentration resulted in a significant decrease in antigenic variation. Although this particular small-molecule ligand is not specific for a particular quadruplex, the result suggests more generally that the future development of more selective quadruplex-targeting agents may have potential in the treatment of a range of pathogenic-induced conditions.

General Reading

An exceptionally large number of reviews on telomeres, telomerase and G-quadruplexes have been published. Many cover specific aspects of these subjects, and some of these are referenced in subsequent chapters. Listed here are those that in particular provide perspectives on therapeutic aspects: the list is not meant to be exclusive and is only limited by space!

De Cian, A., Lacroix, L., Douarre, C., Temime-Smaali, N., Trentesaux, C., Riou, J.-F., et al. 2008. Targeting telomeres and telomerase. Biochimie 90, 131–155.

de Lange, T., Lundblad, V., Blackburn, E.H. (Eds.), 2006. Telomeres (second ed.). Cold Spring Harbor Press, New York.

Kelland, L.R., 2005. Overcoming the immortality of tumour cells by telomere and telomerase based cancer therapeutics—current status and future prospects. Eur. J. Cancer 41, 971–979.

Kipling, D., 1995. The Telomere. Oxford University Press, Oxford, UK.

Mergny, J.-L., Hélène, C., 1998. G-quadruplex DNA: a target for drug design. Nature Med. 4, 1366–1367.

Neidle, S., Balasubramanian, S. (Eds.), 2006. Quadruplex Nucleic Acids. RSC Publishing, Cambridge, UK.

Ouellette, M.M., Wright, W.E., Shay, J.W., 2011. Targeting telomerase-expressing cancer cells. J. Cell. Mol. Med. In press.

Phatak, P., Burger, A.M., 2007. Telomerase and its potential for therapeutic intervention. Brit. J. Pharmacol. 152, 1003–1011.

Rezler, E.M., Bearss, D.J., Hurley, L.H., 2003. Telomere inhibition and telomere disruption as processes for drug targeting. Ann. Rev. Pharmacol. Tox. 43, 359–379.

Shay, J.W., Wright, W.E., 2006. Telomerase therapeutics for cancer: challenges and new directions. Nature Rev. Drug Discov. 5, 577–584.

Wu, Y., Brosh Jr., R.M., 2010. G-quadruplex nucleic acids and human disease. FEBS J. 277, 3470–3488.

References

Autexier, C., Lue, N.F., 2006. The structure and function of telomerase reverse transcriptase. Ann. Rev. Biochem. 75, 493–517.

Bang, I., 1910. Untersuchungen über die guanylsäre. Biochem. Zeitschrift 26, 293–311.

Barma, D.K., Elayadi, A., Falck, J.R., Corey, D.R., 2003. Inhibition of telomerase by BIBR 1532 and related analogues. Bioorg. Med. Chem. Lett. 13, 1333–1336.

Baumann, P., Cech, T.R., 2001. Pot1, the putative telomere end-binding protein in fission yeast and humans. Science 292, 1171–1175.

Blackburn, E.H., Collins, K., 2011. Telomerase: an RNP enzyme synthesizes DNA. Cold Spring Harbor Persp. Biology 3, a003558.

Blackburn, E.H., 1991. Structure and function of telomeres. Nature 350, 569–573.

Bodnar, A.G., Quellette, M., Frolkis, M., Holt, S.E., Chiu, C.-P., Morin, G.B., et al., 1998. Extension of life-span by introduction of telomerase into normal human cells. Science 279, 349–352.

Cahoon, L.A., Seifert, H.S., 2009. An alternative DNA structure is necessary for pilin antigenic variation in *Neisseria gonorrhoeae*. Science 325, 764–767.

Capra, J.A., Paeshke, K., Singh, M., Zakian, V.A., 2010. G-quadruplex DNA sequences are evolutionarily conserved and associated with distinct genomic features in saccharomyces cerevisiae. PLos Comput. Biol. 6, e1000861.

Castelo-Branco, P., Zhang, C., Lipman, T., Fujitani, M., Hansford, L., Clarke, I., et al., 2011. Neural tumor-initiating cells have distinct telomere maintenance and can be safely targeted for telomerase inhibition. Clin. Cancer Res. 17, 111–121.

Cech, T.R., 2000. Life at the end of the chromosome: telomeres and telomerase. Angew. Chem. Int. Ed. Engl. 39, 34–43.

Cech, T.R., 2004. Beginning to understand the end of the chromosome. Cell 116, 273–279.

Cesare, A.J., Reddel, R.R., 2010. Alternative lengthening of telomeres: models, mechanisms and implications. Nat. Rev. Genet. 11, 319–330.

Counter, C.M., Hirte, H.W., Bacchetti, S., Harley, C.B., 1994. Telomerase activity in human ovarian carcinoma. Proc. Natl. Acad. Sci. USA 91, 2900–2904.

d'Adda di Fagagna, F., Teo, S.H., Jackson, S.P., 2004. Functional links between telomeres and proteins of the DNA-damage response. Genes Dev. 18, 1781–1799.

d'Adda-di Fagugna, F., Reaper, P.M., Clay-Farrace, L., Flegler, H., Carr, P., von Zglinicki, T., et al., 2003. A DNA damage checkpoint response in telomere-initiated senescence. Nature 426, 194–198.

Damm, K., Hemmann, U., Garin-Chesa, P., Hauel, N., Kauffmann, I., Priepke, H., et al., 2001. A highly selective telomerase inhibitor limiting human cancer cell proliferation. EMBO J. 20, 6958–6968.

D'Antonio, L. & Bagga, P. 2004 Computational methods for predicting intramolecular G-quadruplexes in nucleotide sequences. 2004 IEEE Computational Systems Bioinformatics Conference, pp. 561–562.

Davis, J.T., 2004. G-quartets 40 years later: from 5′-GMP to molecular biology and supramolecular chemistry. Angew. Chem. Int. Ed. Engl. 43, 668–698.

De Cian, A., Cristofai, G., Reichenbach, P., De Lemos, E., Monchaud, D., Teulade-Fichou, M.-P., et al., 2007. Reevaluation of telomerase inhibition by quadruplex ligands and their mechanisms of action. Proc. Natl. Acad. Sci. USA 104, 17347–17352.

de Lange, T., 2005. Shelterin: the protein complex that shapes and safeguards human telomeres. Genes Develop. 19, 2100–2110.

de Lange, T., 2010. How Shelterin solves the telomere end-replication problem. Cold Spring Harbor Symp. Quant. Biol. 75, 167–177.

Donohue, J., 1956. Hydrogen-bonded helical configurations of polynucleotides. Proc. Natl. Acad. Sci. USA 42, 60–65.

Eddy, J., Maizels, N., 2006. Gene function correlates with potential for G4 DNA formation in the human genome. Nucleic Acids Res. 34, 3887–3896.

Eddy, J., Maizels, N., 2008. Conserved elements with potential to form polymorphic G-quadruplex structures in the first intron of human genes. Nucleic Acids Res. 36, 1321–1330.

Eddy, J., Maizels, N., 2009. Selection for the G4 DNA motif at the 5′ end of human genes. Mol. Carcinogenesis 48, 319–325.

El-Daly, H., Kull, M., Zimmermann, S., Pantic, M., Waller, C.F., Martens, U.M., 2005. Selective cytotoxicity and telomere damage in leukemia cells using the telomerase inhibitor BIBR1532. Blood 105, 1742–1749.

Feldser, D.M., Greider, C.W., 2007. Short telomeres limit tumor progression in vivo by inducing senescence. Cancer Cell 11, 461–469.

Feng, J.L., Funk, W.D., Wang, S.S., Weinrich, S.L., Avilion, A.A., Chiu, C.P., et al., 1995. The RNA component of human telomerase. Science 269, 1236–1241.

Gellert, M., Lipsett, M.N., Davies, D.R., 1962. Helix formation by guanylic acid. Proc. Natl. Acad. Sci. USA 48, 2013–2018.

Gillis, A.J., Schuller, A.P., Skordalakes, E., 2008. Structure of the *Tribolium castaneum* telomerase catalytic subunit TERT. Nature 455, 633–637.

Gomez, D., Mergny, J.-L., Riou, J.-F., 2002. Detection of telomerase inhibitors based on G-quadruplex ligands by a modified telomeric repeat amplification protocol assay. Cancer Res. 62, 3365–3368.

Greider, C.W., Villeponteau, B., 1995. The RNA component of human telomerase. Science 269, 1236–1241.

Greider, C.W., Blackburn, E.H., 1985. Identification of a specific telomere terminal transferase activity in Tetrahymena extracts. Cell 43, 405–413.

Griffith, J.D., Comeau, L., Rosenfield, S., Stansel, R.M., Bianchi, A., Moss, H., et al., 1999. Mammalian telomeres end in a large duplex loop. Cell 97, 503–514.

Hahn, W.C., Counter, C.M., Lundberg, A.S., Beijersbergen, R.L., Brooks, M.W., Weinberg, R.A., 1999a. Creation of human tumour cells with defined genetic elements. Nature 400, 464–468.

Hahn, W.C., Stewart, S.A., Brooks, M.W., York, S.G., Eaton, E., Kurachi, A., et al., 1999b. Inhibition of telomerase limits the growth of human cancer cells. Nat. Med. 5, 1164–1170.

Hanahan, D., Weinberg, R.A., 2000. The hallmarks of cancer. Cell 100, 57–70.

Hanahan, D., Weinberg, R.A., 2011. The hallmarks of cancer. The next generation. Cell 144, 646–674.

Hemann, M.T., Strong, M.A., Hao, L.Y., Greider, C.W., 2001. The shortest telomere, not average telomere length, is critical for cell viability and chromosome stability. Cell 107, 67–77.

Henderson, E., Hardin, C.C., Walk, S.K., Tinoco Jr., I., Blackburn, E.H., 1987. Telomeric DNA oligonucleotides form novel intramolecular structures containing guanine-guanine base pairs. Biochemistry 51, 899–908.

Herbert, B.S., Pitts, A.E., Baker, S.I., Hamilton, S.E., Wright, W.E., Shay, J.W., et al., 1999. Inhibition of human telomerase in immortal human cells leads to progressive telomere shortening and cell death. Proc. Natl. Acad. Sci. USA 96, 14276–14281.

Hershman, S.G., Chen, Q., Lee, J.Y., Kozak, M.L., Yue, P., Wang, L.-S., et al., 2008. Genomic distribution and functional analyses of potential G-quadruplex-forming sequences in Saccharomyces cerevisiae. Nucleic Acids Res. 36, 144–156.

Huppert, J.L., Balasubramanian, S., 2005. Prevalence of quadruplexes in the human genome. Nucleic Acids Res. 33, 2908–2916.

Huppert, J.L., Balasubramanian, S., 2007. G-quadruplexes in promoters throughout the human genome. Nucleic Acids Res. 35, 406–413.

Ijpma, A.S., Greider, C.W., 2003. Short telomeres induce a DNA damage response in *Saccharomyces cerevisiae*. Mol. Biol. Cell 14, 987–1001.

Johnson, J.E., Cao, K., Ryvkin, P., Wang, L.S., Johnson, F.B., 2010. Altered gene expression in the Werner and Bloom syndromes is associated with sequences having G-quadruplex forming potential. Nucleic Acids Res. 38, 1114–1122.

Joseph, I., Tressier, R., Bassett, E., Harley, C., Buseman, C.M., Pattamatta, P., et al., 2010. The telomerase inhibitor Imetelstat depletes cancer stem cells in breast and pancreatic cancer cell lines. Cancer Res. 70, 9494–9504.

Kelland, L.R., 2007. Targeting the limitless replicative potential of cancer: the telomerase/telomere pathway. Clin. Cancer Res. 13, 4960–4963.

Kikin, O., D'Antonio, L., Bagga, P., 2006. QGRS Mapper: a web-based server for predicting G-quadruplexes in nucleotide sequences. Nucleic Acids Res. 34, W676–W682.

Kikin, O., Zappala, Z., D'Antonio, L., Bagga, P.S., 2008. GRSDB2 and GRS_UTRdb: databases of quadruplex forming G-rich sequences in pre-mRNAs and mRNAs. Nucleic Acids Res. 39, D141–D148.

Kim, N.W., Wu, F., 1997. Advances in quantification and characterization of telomerase activity by the telomeric repeat amplification protocol (TRAP). Nucleic Acids Res. 25, 2595–2597.

Kim, N.W., Piatyszek, M.A., Prowse, K.R., Harley, C.B., West, M.D., Ho, P.L., et al., 1994. Specific association of human telomerase activity with immortal cells and cancer. Science 266, 2011–2015.

Kosciolek, B.A., Kalantidis, K., Tabler, M., Rowley, P.T., 2003. Inhibition of telomerase activity in human cancer cells by RNA interference. Mol. Cancer Ther. 2, 209–216.

Law, M.J., Lower, K.M., Voon, H.P., Hughes, J.R., Garrick, D., Viprakasit, V., et al., 2010. ATR-X syndrome protein targets tandem repeats and influences allele-specific expression in a size-dependent manner. Cell 143, 367–378.

Lei, M., Podell, E.M., Cech, T.R., 2004. Structure of human POT1 bound to telomeric single-stranded DNA provides a model for chromosome end-protection. Nat. Struct. Mol. Biol. 11, 1223–1229.

Li, G.-Z., Eller, M.S., Firoozabadi, R., Gilchrest, B.A., 2003. Evidence that exposure of the telomere 3′ overhang sequence induces senescence. Proc. Natl. Acad. Sci. USA 100, 527–531.

Li, J.L., Harrison, R.J., Reszka, A.P., Brosh Jr., R.M., Bohr, V.A., Neidle, S., et al., 2001. Inhibition of the Bloom's and Werner's syndrome helicases by G-quadruplex interacting ligands. Biochemistry 40, 15194–15202.

Lipps, H.J., Rhodes, D., 2009. G-quadruplex structures: *in vivo* evidence and function. Trends Cell Biol. 19, 414–422.

London, T.B., Barber, L.J., Mosedale, G., Kelly, G.P., Balasubramanian, S., Hickson, I.D., et al., 2008. FANCJ is a structure-specific DNA helicase associated with the maintenance of genomic G/C tracts. J. Biol. Chem. 283, 36132–36139.

Maizels, N., 2006. Dynamic roles for G4 DNA in the biology of eukaryotic cells. Nat. Struct. Mol. Biol. 13, 1055–1059.

Marian, C.O., Cho, S.K., Mcellin, B.M., Maher, E.A., Hatanpaa, K.J., Madden, C.J., et al., 2010. The telomerase antagonist Imetelstat efficiently targets glioblastoma tumor-initiating cells leading to decreased proliferation and tumor growth. Clin. Cancer Res. 16, 154–163.

Mason, M., Schuller, A.P., Skordalakes, E., 2010. Telomerase structure function. Curr. Opin. Struct. Biol. 21, 1–9.

Melana, S.M., Holland, J.F., Pogo, B.G., 1998. Inhibition of cell growth and telomerase activity of breast cancer cells in vitro by 3′-azido-3′-deoxythymidine. Clin. Cancer Res. 4, 693–696.

Mergny, J.-L., Riou, J.-F., Maillet, P., Teulade-Fichou, M.-P., Gilson, E., 2002. Natural and pharmacological regulation of telomerase. Nucleic Acids Res. 30, 839–865.

Meyerson, M., Counter, C.M., Eaton, E.N., Ellisen, L.W., Steiner, P., Caddle, S.D., et al., 1997. hEST2, the putative human telomerase catalytic subunit gene, is up-regulated in tumor cells and during immortalization. Cell 90, 785–795.

Meyne, J., Ratliff, R.L., Moyzis, R.K., 1989. Conservation of the human telomere sequence (TTAGGG)$_n$ among vertebrates. Proc. Natl. Acad. Sci. USA 86, 7049–7053.

Mitchell, M., Gillis, A., Futahashi, M., Fujiwara, H., Skordalakes, E., 2010. Structural basis for telomerase catalytic subunit TERT binding to RNA template and telomeric DNA. Nat. Struct. Mol. Biol. 17, 513–518.

Moyzis, R.K., Buckingham, J.M., Cram, I.S., Dani, M., Deaven, L.L., Jones, M.D., et al., 1988. A highly conserved repetitive DNA sequence $(TTAGGG)_n$ present at the telomeres of human chromosomes. Proc. Natl. Acad. Sci. USA 85, 6622–6626.

Müller, S., Kumari, S., Rodriguez, R., Balasubramanian, S., 2011. Small-molecule-mediated G-quadruplex isolation from human cells. Nat. Chem. 2, 1095–1098.

Murchie, A.I., Lilley, D.M., 1992. Retinoblastoma susceptibility genes contain 5′ sequences with a high propensity to form guanine-tetrad structures. Nucleic Acids Res. 20, 49–53.

Nakamura, T.M., Morin, G.B., Chapman, K.B., Weinrich, S.L., Andrews, W.H., Lingner, J., et al., 1997. Telomerase catalytic subunit homologs from fission yeast and human. Science 277, 955–959.

Oganesian, L., Bryan, T.M., 2007. Physiological relevance of telomeric G-quadruplex formation: a potential drug target. BioEssays 29, 155–165.

Paeschke, K., Juranek, S., Simonsson, T., Hempel, A., Rhodes, D., Lipps, H.J., 2008. Telomerase recruitment by the telomere end binding protein-beta facilitates G-quadruplex DNA unfolding in ciliates. Nat. Struct. Mol. Biol. 15, 598–604.

Pascolo, E., Wenz, C., Lingner, J., Hauel, N., Priepke, H., Kauffmann, I., et al., 2002. Mechanism of human telomerase inhibition by BIBR1532, a synthetic, non-nucleosidic drug candidate. J. Biol. Chem. 277, 15566–15572.

Ralph, R.K., Connors, W.J., Khorana, H.G., 1962. Secondary structure and aggregation in deoxyguanosine oligonucleotides. J. Amer. Chem. Soc. 84, 2265–2266.

Rawal, P., Kummarasetti, V.B.R., Ravindran, J., Kurmar, N., Halder, K., Sharma, R., et al., 2006. Genome-wide prediction of G4 DNA as regulatory motifs: role in Escherichia coli global regulation. Genome Res. 16, 644–655.

Reaper, P.M., d'Adda di Fagagna, F., Jackson, S.P., 2004. Activation of the DNA damage response by telomere attrition: a passage to cellular senescence. Cell Cycle 3, 543–546.

Reed, J., Gunaratnam, M., Beltran, M., Reszka, A.P., Vilar, R., Neidle, S., 2008. TRAP-LIG, a modified telomere repeat amplification protocol assay to quantitate telomerase inhibition by small molecules. Anal. Biochem. 380, 99–105.

Röth, A., Harley, C.B., Baerlocher, G.M., 2010. Imetelstat (GRN163L)-telomerase-based cancer therapy. Recent Results Cancer Res. 184, 221–234.

Ryvkin, P., Hershman, S.G., Wang, L.S., Johnson, F.B., 2010. Computational approaches to the detection and analysis of sequences with intramolecular G-quadruplex forming potential. Methods Mol. Biol. 608, 39–50.

Schaffitzel, C., Berger, I., Postberg, J., Hanes, J., Lipps, H.J., Plückthun, A., 2001. In vitro generated antibodies specific for telomeric guanine-quadruplex DNA react with *Stylonychia lemnae* macronuclei. Proc. Natl. Acad. Sci. USA 98, 8572–8577.

Schalch, T., Duda, S., Sargent, D.F., Richmond, T.J., 2005. X-ray structure of a tetranucleosome and its implications for the chromatin fibre. Nature 436, 138–141.

Sen, D., Gilbert, W., 1988. Formation of parallel four-stranded complexes by guanine-rich motifs in DNA and its implications for meiosis. Nature 334, 364–366.

Shay, J.W., Bacchetti, S., 1997. A survey of telomerase activity in human cancer. Eur. J. Cancer 33, 787–791.

Shay, J.W., Wright, W.E., 2010. Telomeres and telomerase in cancer and normal stem cells. FEBS Lett. 584, 3619–3625.

Shay, J.W., Wright, W.E., 2010. Telomerase: a target for cancer therapeutics. Cancer Cell 2, 257–265.

Simonsson, T., 2001. G-quadruplex DNA structures—variations on a theme. Biol. Chem. 382, 621–628.

Simonsson, T., Pecinka, P., Kubista, M., 1998. DNA tetraplex formation in the control region of c-myc. Nucleic Acids Res. 26, 1167–1172.

Soldatenkov, V.A., Vetcher, A.A., Duka, T., Ladame, S., 2008. First evidence of a functional interaction between DNA quadruplexes and poly(ADP-ribose) polymerase-1. ACS Chem. Biol. 3, 214–219.

Stewart, S.A., Weinberg, R.A., 2006. Telomeres: cancer to human aging. Ann. Rev. Cell Develop. Biol. 22, 531–557.

Strahl, C., Blackburn, E.H., 1994. The effects of nucleoside analogs on telomerase and telomeres in Tetrahymena. Nucleic Acids Res. 22, 893–900.

Strahl, C., Blackburn, E.H., 1996. Effects of reverse transcriptase inhibitors on telomere length and telomerase activity in two immortalized human cell lines. Mol. Cell Biol. 16, 53–65.

Sun, D., Hurley, L.H., Von Hoff, D.D., 1998. Telomerase assay using biotinylated-primer extension and magnetic separation of the products. Biotechniques 25, 1046–1051.

Sun, D., Thompson, B., Cathers, B.E., Salazar, M., Kerwin, S.M., Trent, J.O., et al., 1997. Inhibition of human telomerase by a G-quadruplex-interactive compound. J. Med. Chem. 40, 2113–2116.

Sun, H., Karow, J.K., Hickson, I.D., Maizels, N., 1998. The Blooms' syndrome helicase unwinds G4 DNA. J. Biol. Chem. 273, 27587–27592.

Sun, H., Yabuki, A., Maizels, N., 2002. A human nuclease specific for G4 DNA. Proc. Natl. Acad. Sci. USA 98, 12444–12449.

Sundquist, W.I., Klug, A., 1989. Telomeric DNA dimerizes by formation of guanine tetrads between hairpin loops. Nature 342, 825–829.

Todd, A.K., Neidle, S., 2011. Mapping the sequences of potential guanine quadruplex motifs. Nucleic Acids Res. In press

Todd, A.K., 2007. Bioinformatics approaches to quadruplex sequence location. Methods 43, 246–251.

Todd, A.K., Johnston, M., Neidle, S., 2005. Highly prevalent putative quadruplex sequence motifs in human DNA. Nucleic Acids Res. 33, 2901–2907.

Wang, Q., Liu, J.-Q., Chen, Z., Zheng, K.-W., Chen, C.-Y., Hao, Y.-H., et al., 2011. G-quadruplex formation at the 3′ end of telomere DNA inhibits its extension by telomerase, polymerase and unwinding by helicase. Nucleic Acids Res. In press

Wright, W.E., Shay, J.W., Piatyszek, M.A., 1995. Modifications of a telomeric repeat amplification protocol (TRAP) result in increased reliability, linearity and sensitivity. Nucleic Acids Res. 23, 3794–3795.

Wright, W.E., Tesmer, V.M., Huffman, K.E., Levene, S.D., Shay, J.W., 1997. Normal human chromosomes have long G-rich telomeric overhangs at one end. Genes Dev. 11, 2801–2809.

Wu, Y., Shin-ya, K., Brosh Jr., R.M., 2008. FANCJ helicase defective in Fanconia anemia and breast cancer unwinds G-quadruplex DNA to defend genomic stability. Mol. Cell. Biol. 28, 4116–4128.

Wyatt, H.D., West, S.C., Beattie, T.L., 2010. InTERTpreting telomerase structure and function. Nucleic Acids Res. 38, 5609–5622.

Yadav, V.K., Abraham, J.K., Mani, P., Kulshrestha, R., Chowdhury, S., 2008. QuadBase: genome-wide database of G4 DNA occurrence and conservation in human, chimpanzee, mouse and rat promoters and 146 microbes. Nucleic Acids Res. 36, D381–D385.

Zahler, A.M., Williamson, J.R., Cech, T.R., Prescott, D.M., 1991. Inhibition of telomerase by G-quartet DNA structures. Nature 350, 718–750.

Zaug, A.J. Podell, E.R. & Cech, T.R. (2005) Human POT1 disrupts telomeric G-quadruplexes allowing telomerase extension in vitro. PNAs 102, 10864–10869.

Zhang, L., Mar, V., Zhou, W., Harrington, L., Robinson, M.O., 1999. Telomere shortening and apoptosis in telomerase-inhibited human tumor cells. Genes Dev. 13, 2388–2399.

Zhang, R., Lin, Y., Zhang, C.-T., 2007. Greglist: a database listing potential G-quadruplex regulated genes. Nucleic Acids Res. 36, D372–D376.

2

DNA and RNA Quadruplex Structures

The G-Quartet—The Cement that Holds Quadruplex Structures Together

The core motif of all quadruplex nucleic acids, the G-quartet, has been introduced in Chapter 1. This chapter starts by examining the arrangement in greater detail. The G-quartet involves two edges (the Watson-Crick and the Hoogsteen) of each of the four guanine bases (Gellert, Lipsett & Davies, 1962), with each base accepting two (to atoms O6 and N7), and donating two hydrogen bonds (from atoms N2 and N3), so that the overall arrangement (Figure 2–1a) has a total of eight hydrogen bonds (four N1-H...O6 and four N2-H...N7 hydrogen bonds). X-ray crystallographic studies on short-length oligonucleotide quadruplex structures have confirmed and extended the earlier proposals for the G-quartet arrangement, and although only very few of these single-crystal analyses are at true atomic resolution (Table 2–1), cumulatively they provide a reliable set of hydrogen-bond distances and angles. Nitrogen–oxygen hydrogen-bond distances in G-quartets within quadruplex crystal and NMR structures range between 2.7 and 3.0 Å, although the most recent high-resolution crystal structures find that all eight hydrogen bonds are in a narrower range of 2.85 to 2.95 Å (see, for example, the $d[TGGGGT]_4$ quadruplex structure PDB id 352d at 0.95 Å resolution: Phillips *et al.*, 1997; also the *Oxytricha nova* $d[G_4T_4G_4]$ quadruplex ligand complex structure PDB id 3NYP at 1.18 Å resolution: Campbell *et al.*, 2011; Figure 2–1b). Contrary to the idealized geometry assumed in the earlier models built from the fiber diffraction studies, individual guanine bases in a G-quartet are commonly found even in the highest-resolution quadruplex crystal structures to have up to 10–15° of inclination with respect to each other. The early molecular dynamics simulations on quadruplexes (Špačková, Berger & Šponer, 1999, 2001) sometimes reported that additional three-center bifurcated hydrogen-bond arrangements can be formed involving the O6 atom on one guanine and the hydrogen atoms attached to both N1 and N2 atoms on the adjacent guanine. However, this abnormal feature has not been observed to date in high-resolution quadruplex crystal structures or inferred from NMR models, and is likely to be an artifact of the earlier force fields used. In, for example, the more recent high-resolution crystal structures, the N2...O6 distance is consistently at ~3.4 Å, i.e. exactly at van der Waals separation. Also more recent improvements to, in particular, backbone parameterizations for the AMBER nucleic acid force field, coupled with greater

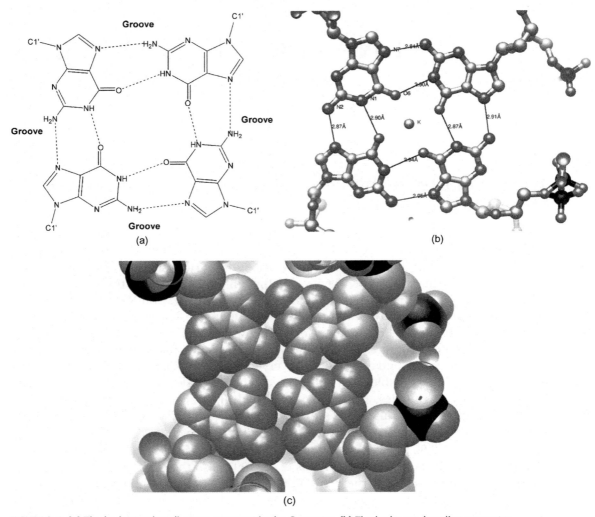

FIGURE 2–1 (a) The hydrogen-bonding arrangement in the G-quartet. **(b)** The hydrogen-bonding geometry observed in one of the G-quartets in the *Oxytricha nova* d[G₄T₄G₄] quadruplex ligand complex crystal structure PDB id 3NYP at 1.18Å resolution (Campbell *et al.*, 2011). The distances are accurate to ±0.02 Å. **(c)** Space-filling representation of the G-quartet, viewed onto the plane of the quartet.

attention to long-range electrostatics, have led to increasingly reliable molecular dynamics simulations of quadruplex structures (Šponer & Špačková, 2007). These are discussed in Section 2.3 below in more detail. A quantum mechanics density functional study (Meyer, Brandl & Sühnel, 2001) suggested that the precise hydrogen-bonding arrangement depends on the planarity of the G-quartet, with the classical Hoogsteen arrangement more favored by a slightly non-planar arrangement. However, the small energy

Table 2–1   Tetramolecular and bimolecular quadruplex structures based on non-mammalian nucleic acid sequences that have been reported, and are available in the Protein Data Bank. Loop types are designated by D (diagonal), L (lateral) and P (propeller). Other quadruplex structures are detailed in later chapters

Sequence	Method	Topology	Loops	PDB id	Reference
d[TG$_4$T] + Ca^{2+}/Na$^+$	X-ray	parallel	none	352D	Phillips *et al.*, 1997
d[TG$_4$T] + Ca^{2+}/Na$^+$	X-ray	parallel	none	2GW0	Lee *et al.*, 2007
d[TG$_4$T] + Tl$^+$/Na$^+$	X-ray	parallel	none	1S45	Cáceres *et al.*, 2004
d[TG$_4$T] + Li$^+$	X-ray	parallel	none	2O4F	Creze *et al.*, 2007
d[G$_4$T$_4$G$_4$] + K$^+$	X-ray	anti-parallel	2 × D	1JPQ, 1JRN	Haider, Parkinson & Neidle, 2003
d[G$_4$T$_4$G$_4$] + telomeric protein	X-ray	anti-parallel	2 × D	1JB7	Horvath & Schultz, 2001
d[G$_4$T$_4$G$_4$] + K$^+$	NMR	anti-parallel	2 × D	1K4X	Schultze *et al.*, 1999
d[G$_4$T$_4$G$_4$] + Li$^+$	X-ray	anti-parallel	2 × D	2HBN	Gill, Strobel & Loria, 2006
d[G$_4$T$_4$G$_4$] + Na$^+$	NMR	anti-parallel	2 × D	156D	Schultze, Smith & Feigon, 1994
d[G$_4$T$_4$G$_4$] + Li$^+$	NMR	anti-parallel	2 × D	2AKG	Gill, Strobel & Loria, 2005
d[G$_3$T$_4$G$_3$] + Na$^+$	NMR	anti-parallel	2 × D	1FQP	Keniry *et al.*, 1995
d[G$_4$T$_4$G$_3$] + Na$^+$	NMR	anti-parallel	2 × D	1LVS	Čmugelj, Hud & Plavec, 2002
d[G$_3$T$_4$G$_4$] + Na$^+$	NMR	anti-parallel	1 × D; 1 × L	1U64	Šket, Čmugelj & Plavec, 2003
d[G$_4$T$_3$G$_4$] + Na$^+$	X-ray	anti-parallel	2 × L	2AVH	Hazel, Parkinson & Neidle, 2006
d[G$_4$(T$_4$G$_4$)$_3$] + Na$^+$	NMR	anti-parallel	1 × D; 2 × L	201D	Wang & Patel, 1995

differences are sensitive to the basis set used, and it is tempting to conclude that in this instance experiment is in advance of theory. The G-quartet has approximately twice the surface area of a Watson-Crick base pair (Figure 2–1c). The C1′–C1′ distance (measured in the high-resolution 1.18Å quadruplex crystal structure: PDB id 3NYP) varies between 11.5 and 11.7 Å, so that the overall surface area of a G-quartet is ~135 Å^2.

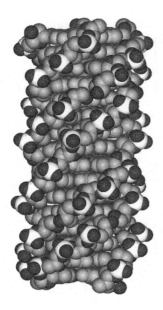

FIGURE 2–2 Representation of the poly (rG) four-fold helix, with the phosphate groups shown by highlighted shading.

Four-stranded Helix Models

Fiber-diffraction studies subsequent to the original analysis of Gellert *et al.* (Zimmerman, 1976) have confirmed their four-stranded helical model (Figure 2–2) with stacked G-quartets for the gel structure of 5′-guanosine monophosphate, although an alternative arrangement was subsequently proposed (Walmsley & Burnett, 1999), based on a solution NMR study, in which the hydrogen-bonded guanines in a G-quartet are no longer even approximately coplanar but each individual base plane is twisted with respect to the next so as to form a continuous helical arrangement. A more recent NMR structure analysis, also in solution (Wu & Kwan, 2009) has derived a detailed model for 5′-guanosine monophosphate that is closely analogous to the original fiber-diffraction model, and is consistent with a four-fold right-handed helix having alternating C2′-*endo* and C3′-*endo* ribose sugar puckers (for definitions see Chapter 10).

Fiber diffraction studies have also been undertaken on polynucleotides comprising solely guanosine repeats (Arnott, Chandresakaran & Marttila, 1974; Zimmerman, Cohen & Davies, 1976), and extensive biophysical studies have been reported on them (Thiele & Guschelbauer, 1971; Howard, Frazier & Miles, 1977). These analyses all conclude that poly (rG) forms a four-stranded continuous helical structure (Figure 2–2) whose dimensions are fully consistent with the structural motif being the G-quartet. The arrangement, as defined from fiber-diffraction analysis, is of a right-handed helix with an axial translation per G-quartet of 3.4 Å, a rotation per G-quartet of 31° and C2′-*endo* sugar puckers (Arnott, Chandrasekaran & Marttila, 1974). The G-quartets are only slightly tilted with respect to the helix axis so that the overall appearance is remarkably similar to that of a somewhat

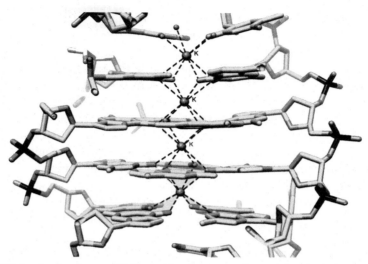

FIGURE 2–3 Plot of the potassium ions and their coordination to guanine and thymine bases, as observed in the central channel of the *Oxytricha nova* d[G₄T₄G₄] quadruplex ligand complex crystal structure PDB id 3NYP at 1.18 Å resolution (Campbell *et al.*, 2011).

wider B-DNA double helix—it is necessary to examine the poly (rG) helix closely to realize that it is a four-fold helix and not a double helix!

The Role of Metal Ions

The early studies on guanosine self-association had an explicit realization that there is a requirement for particular metal ions in order to stabilize the resulting structures, which is distinct from the requirements of duplex DNA. The latter are stabilized by Mg^{2+} ions, which do not stabilize four-stranded G-quartet structures. Early models for four-fold helices and oligonucleotide structures based on the G-quartet motif (for example, Howard & Miles, 1982) suggested the existence of a central channel into which water molecules or cations could be placed. This interpretation was subsequently confirmed (see, for example, Oka & Thomas, 1987; Sen & Gilbert, 1988; Hardin *et al.*, 1991; Williamson *et al.*, 1989) and extended by a large number of biophysical and structural studies on G-quartet-containing nucleic acid sequences (reviewed by Hud & Plavec, 2006). Of particular biological importance have been the findings that the stability of these four-stranded structures depends on the presence of physiological concentrations of the alkali metals sodium or potassium with the order of stabilizing ability being $K^+ > Na^+$. The precise location of an ion in the channel depends on its ionic radius. Thus K^+ ions are too large to be accommodated in the plane of a G-quartet, but reside midway between successive G-quartets (Figure 2–3), 2.8–2.9 Å from each of the eight O6 carbonyl atoms of the guanines, forming a bipyramidal antiprismatic arrangement. K^+–K^+ distances along the channel are generally 3.4–3.5 Å.

Na$^+$ ions are small enough to be coordinated in-plane with all four guanine bases in a G-quartet. This arrangement is seen for some of the bound Na$^+$ ions in the high-resolution d[TGGGGT]$_4$ tetramolecular quadruplex crystal structure (Phillips *et al.*, 1997), where the Na$^+$ ions are not evenly distributed along the channel, having Na$^+$–Na$^+$ distances varying between 3.4 and 4.3 Å. At the ends of the channel the Na$^+$ coordination is almost square-planar with the four close O6 guanine neighbors (Na$^+$–O6 separations are ~2.3 Å). However, they are not constrained always to adopt this arrangement and their coordination within the same structure elsewhere along the channel is the same bipyramidal antiprismatic arrangement as is found for K$^+$ ions (Figure 2–4), albeit not always symmetrically arranged and with Na$^+$–O6 distances variable within the range 2.6–3.0 Å. The range of positions found experimentally for Na$^+$ ions along the channel suggests that the energy barrier to their movement for small ions is low, in accordance with both theory (for example, van Mourik & Dingley, 2005) and experiment (Sket & Plavec, 2010).

Several other ions have been reported to stabilize quadruplexes, including those of divalent heavier metals such as rubidium and caesium (Rb$^+$, Cs$^+$: Ida, Kwan & Wu, 2007; Marincola *et al.*, 2009), strontium (Sr^{3+}: Chen, 1992; Pedroso *et al.*, 2007), thallium (Tl$^+$: Gill, Strobel & Loria, 2006), calcium (Ca^{2+}: Lee *et al.*, 2007; Ida & Wu, 2008), lead (Pb^{2+}: Smirnov & Shafer, 2000; Smirnov *et al.*, 2002), barium (Ba^{2+}) and various lanthanide metal ions (Kwan, She & Wu, 2007), though all to a lesser extent than K$^+$. Crystallographic data are available on the tetramolecular d[TGGGGT] quadruplex (see Section 2.4) with both discrete Ca^{2+} and Na$^+$ ions observed in the channel (Lee *et al.*, 2007); surprisingly the Ca^{2+}–O6 coordination is bipyramidal antiprismatic, closely resembling K$^+$

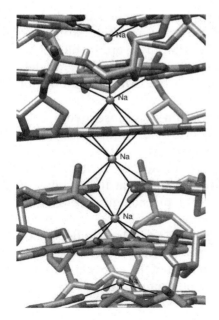

FIGURE 2–4 Plot of the sodium ions and their coordination to guanine O6 atoms, observed in the central channel of the d[TGGGGT] crystal structure (Phillips *et al.*, 1997), highlighting the different positions of the Na$^+$ ions and their consequent variation in coordination distances.

arrangements. This suggests that heavy-atom divalent metals are relatively immobilized in the ion channel and could be useful in phasing crystal structures of quadruplexes and their complexes in the future, although there have not been any reports of this to date. The ions are always located in crystal structures and indeed when not found in electron density maps their absence is a cause for concern about the correctness of the structure. Metal ion coordination patterns observed in X-ray crystallographic analyses generally concur with findings from several multinuclear NMR studies. These latter have been able to directly locate Na^+, K^+, Rb^+ and Ca^{2+} ions (Ida & Wu, 2008). The presence of particular ions in the channel can also change quadruplex conformation and fold (Sen & Gilbert, 1990; Balagurumoorthy & Brahmachari, 1994), an effect that has been most extensively documented for human telomeric quadruplexes (see Chapter 4). For example, Sr^{2+} ions have been reported to induce a parallel fold in human telomeric quadruplexes (Pedroso *et al.*, 2007).

The unique electrostatic features of quadruplexes, with a highly anionic exterior and the polar interior channel filled with metal ions, have made molecular dynamics simulations of DNA quadruplexes so much more challenging than those of duplex DNA (Špačková, Berger & Šponer, 1999, 2001). The advent of newer force-field parameterizations for nucleic acids (for example, by Pérez *et al.*, 2007, improving the α/γ backbone torsion angle rotational representation in the widely used AMBER force field), coupled with use of the particle-mesh Ewald treatment of electrostatics, has enabled significantly more stable simulations of quadruplexes to be achieved, but such calculations are still some way from being able to predict complete folding pathways for these structures (Šponer & Špačková, 2007).

Overview of Quadruplex Structural Features

Quadruplexes can be formed from one, two or four separate strands of DNA or RNA (Figure 2–5a–d) and can display a wide variety of topologies, which are in part a consequence of various possible combinations of strand direction, together with variations in loop size and sequence. In principle, a quadruplex could be formed from three separate strands, although it is not clear that this would have biological relevance.

Quadruplexes can be defined in general terms as structures formed by a core of at least two stacked G-quartets, which are held together by loops arising from the intervening mixed-sequence nucleotides that are not usually involved in the quartets themselves. The combination of the number of stacked G-quartets, the polarity of the strands, and the location and length of the loops would be expected to lead to a plurality of G-quadruplex structural types, as indeed is found experimentally.

Potential **unimolecular** (i.e. intramolecular) G-quadruplex forming sequences can be described as being formed by repetition of a **G-tract** sequence motif, together with connector sequences, all within a single run of sequence (Figure 2.5d):

$$G_m X_n G_m X_o G_m X_p G_m$$

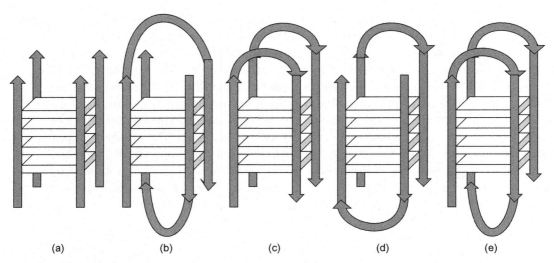

FIGURE 2–5 The different categories of quadruplex arrangements. In each instance four stacked G-quartets are shown as rectangular solids and strand directions are indicated by arrows: **(a)** tetramolecular with four separate strands, **(b)** one bimolecular arrangement, showing two diagonal loops, **(c)** another bimolecular arrangement showing two lateral loops, oriented head to head, **(d)** a third bimolecular arrangement, this time with the lateral loops arranged head to tail, **(e)** an intramolecular quadruplex, with two lateral and one diagonal loop.

where m is the number of G residues in each short G-tract, which are usually directly involved in G-quartet interactions. X_n, X_o and X_p can be any combination of residues, including G, forming the connecting intervening sequences. This simple notation has all the G-tracts of equal length. However, this is not a necessary condition since if one or more of the short G tracts is longer than the others, some G residues will be located in the connector regions. Also, the presence of G residues within connecting sequences can give rise to more complex arrangements in which such individual G residues become inserted into the G-quartet stack and displace the G-tract G residues.

The assumption that all G-tracts within a quadruplex sequence are identical is normally true for vertebrate telomeric sequences, but is frequently not the case for nontelomeric genomic (as shown by bioinformatics surveys of whole genome sequences, outlined in Chapter 1) or RNA quadruplex sequences (Chapter 8), or even for telomeric sequences in some lower eukaryotics.

In principle **tetramolecular** (tetrameric) and **bimolecular** (dimeric) quadruplexes can each be formed from the association of non-equivalent sequences, although very few quadruplexes with such features have yet been studied in detail. Thus almost all bimolecular quadruplexes reported to date are formed by the association of two identical sequences $X_nG_mX_oG_mX_p$, where n and p may or may not be zero. Tetramolecular quadruplexes may be formed by four $X_nG_mX_o$ strands associating together, where X_n and X_o may be terminal sequences of zero or non-zero length.

The connecting sequences between successive G-tracts can link stacked G-quartets in a number of distinct ways (Esposito *et al.*, 2007), so that a wide variety of quadruplex

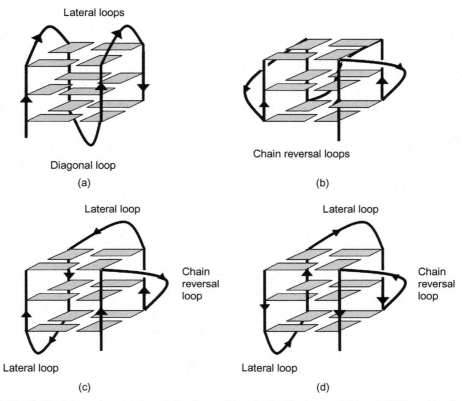

FIGURE 2–6(a–d) The four major experimentally observed topologies (for human telomeric DNA quadruplex sequences—see Chapter 3 for more detail on these) for intramolecular quadruplexes.

topologies can be formed (Figure 2–6). These connectors can form loops (Figure 2–7), which are denoted as **diagonal**, **lateral** (or edgewise) or **propeller** (or chain-reversal). The particular type formed is dependent on the number of G-quartets comprising the stem of a quadruplex, on loop length and sequence and sometimes on the nature of the alkali metal ion. Propeller loops can only be formed by connecting two strands in the same, i.e. parallel, orientation, linking the bottom G-quartet with the top G-quartet in a G-stack whereas diagonal and lateral loops connect chains in opposing, anti-parallel orientations. Lateral loops join adjacent G-strands. Two lateral loops can be located either on the same (Figures 2–5d, 2–7b) or opposite faces of a quadruplex (Figure 2–5c), corresponding to head-to-head or head-to-tail, respectively, when involved in bimolecular quadruplexes. Strand polarities can vary, with one being a head-to-tail lateral loop dimer in which all adjacent strands are anti-parallel, and the other is a head-to-head hairpin quadruplex with one adjacent strand parallel and the other is anti-parallel. The second type of anti-parallel loop, the diagonal loop, joins opposite G-strands. In this instance the directionalities of adjacent strands must alternate between parallel and anti-parallel, and are arranged around a core of stacked G-quartets. All-parallel quadruplexes have all

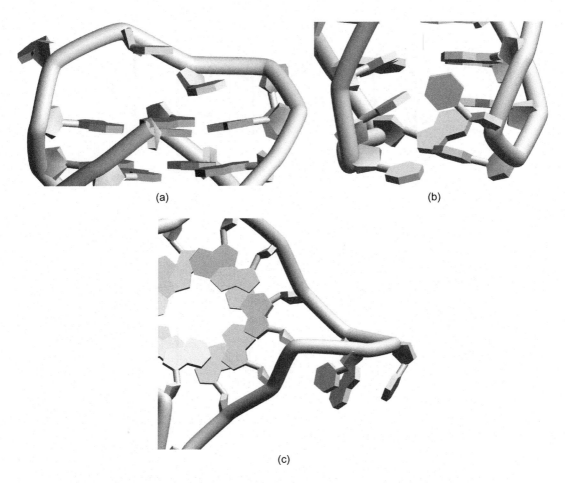

(a)

(b)

(c)

FIGURE 2–7 Views in cartoon form of **(a)** a diagonal loop, **(b)** two lateral loops side by side in a head-to-head arrangement, **(c)** a propeller loop. The first two are necessarily arranged over a G-quartet whereas the third type is splayed out from the G-quartet core.

guanine glycosidic angles in an *anti* conformation (see Chapter 10) whereas anti-parallel quadruplexes normally have equal numbers of *syn* and *anti* guanines, arranged in a way that is particular for a given topology. All quadruplex structures have four grooves, defined as the cavities bounded by the phosphodiester backbones (Figure 2–1). Groove dimensions are variable, and depend on overall topology and the nature of the loops (further details of groove dimensions are given in Chapter 9). Grooves in quadruplexes with only lateral or diagonal loops are structurally simple, and the walls of these grooves are bounded by monotonic sugar phosphodiester groups. By contrast, grooves that incorporate propeller loops have more complex structural features that reflect the insertion of the variable-sequence loops into the grooves.

FIGURE 2–8 Cartoon representation of the d[TGGGGT] tetramolecular quadruplex in one of the crystal structures of this sequence (Lee *et al.*, 2007).

Representative Quadruplex Structures

The crystal and NMR structures of a considerable number of quadruplexes with sequences from mammalian and non-mammalian telomeres, genomic regions or from purely synthetic sources have been determined (see reviews by Phan, Kuryavyi & Patel, 2006; Qin & Hurley, 2008; Neidle, 2009; Phan, 2010) and coordinates for them are available in the PDB. We focus in this chapter, not on the complete database of quadruplexes but instead on just two structurally simple types, tetramolecular and bimolecular. Both are of historic interest, but also illustrate several general features, and continue to be used as paradigms for issues such as ion and ligand binding, so they will be described here. Quadruplexes that are constructed from human genomic sequences are discussed in subsequent chapters.

The yeast telomeric sequence d(TGGGGT) forms a four-stranded tetramolecular quadruplex with all strands parallel to each other and all guanosine glycosidic angles in an *anti* conformation (Table 2–1). This arrangement was first visualized by NMR (Aboul-ela *et al.*, 1994), and the same arrangement was determined in greater detail in a high-resolution crystal structure of the Na^+-containing quadruplex, ultimately at 0.95 Å resolution (Laughlan *et al.*, 1994; Phillips *et al.*, 1997). Both techniques concur in showing the same overall arrangement and there are no discernible differences between the structural core of four stacked G-quartets, which are arranged in a right-handed helical manner with the terminal thymine nucleotides free to adopt a wide range of conformations (Figure 2–8). The arrangement of metal ions in the central channel of this quadruplex has been described in Section 2.3 above. Crystallographic attempts to locate ions other than K^+, Na^+ or Ca^{2+} in the channel have only been modestly successful. Thus Tl^+ ions, which would in principle be useful for phasing crystal structures, have been found to occupy only 15% of one Na^+ site in the d[TGGGGT] quadruplex (Cáceres *et al.*, 2004). The Li^+ ion

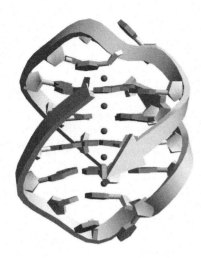

FIGURE 2–9 Cartoon representation of the native *Oxytricha nova* d[$G_4T_4G_4$] quadruplex crystal structure. The K$^+$ ions are shown, and strand directions are indicated by arrows (Haider, Parkinson & Neidle, 2003).

does not occupy a channel site at all, at least in this quadruplex, probably on account of its exceptionally small size (Creze *et al.*, 2007).

In striking contrast two very distinct topologies were initially reported for structures determined by crystallography and NMR for the bimolecular quadruplex formed by the association of two d($G_4T_4G_4$) strands of the telomeric DNA sequence from the *Oxytricha nova* organism. This, for d($G_4T_4G_4$) crystallized from K$^+$ solution, was the first crystal structure of any quadruplex (Kang *et al.*, 1992), and was reported as being an anti-parallel hairpin dimer with two lateral loops. The NMR structure analysis, in Na$^+$ solution, revealed a very different anti-parallel topology, with two diagonal loops (Smith & Feigon, 1992; Schultze, Smith & Feigon, 1994), so that the two halves of the bimolecular quadruplex are intimately inter-digitated together (Figure 2–9), with a central core of four G-quartets and a pattern of *syn-anti-syn-anti-* glycosidic angles along each strand. A four-thymine diagonal loop is positioned over each G-quartet end. It was initially suggested that the differences between the two arrangements was a consequence of the differences in metal ion, but extensive NMR studies on this sequence in K$^+$, Na$^+$ and NH$_4^+$ ion solution (Hud, Schultze & Feigon, 1998; Schultze *et al.*, 1999) demonstrated that there is complete conservation of the structure, at least in dilute solution. For some, while the belief arose that the crystallographic-observed topology was an artefact and that packing forces in quadruplex crystals induce the d($G_4T_4G_4$) sequence into a topology particular to this crystal structure packing arrangement, two subsequent crystal structures have demonstrated that this is not the case, with the implication that the earlier structural assignment (Kang *et al.*, 1992) very unusually does not have the appropriate phosphodiester backbone chain tracing. The d($G_4T_4G_4$) sequence was co-crystallized with an *Oxytricha nova* telomeric protein in the presence of Na$^+$ ions, and the resulting crystal structure at 1.86 Å resolution clearly shows the anti-parallel quadruplex having the identical topology

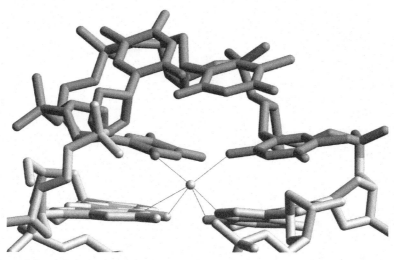

FIGURE 2–10 View of the T4 loop in the *Oxytricha nova* d[G₄T₄G₄] quadruplex crystal structure, with thymine residues shaded gray. The K⁺ ion at the end of the ion channel is shown coordinating to the thymine bases as well as to the guanines of the terminal G-quartet.

as in the NMR experiments, i.e. with diagonal loops (Horvath & Schultz, 2001). Four Na⁺ ions were located in the central channel of the quadruplex. The determination of the crystal structure (Figure 2–9) of the K⁺-containing d(G₄T₄G₄) quadruplex in two distinct crystal forms (Haider, Parkinson & Neidle, 2003) yet both displaying the anti-parallel diagonal loop topology, further confirmed that there is a single stable topology for this sequence that is invariant with respect to solution or crystalline state, or to the nature of the metal ion. The thymine bases in this structure are well ordered and are involved in several interactions: thymine–thymine stacking, thymine–thymine pairing, thymine–G-quartet stacking and thymine–potassium coordination (Figure 2–10). The intramolecular quadruplex d[G₄(T₄G₄)₃] formed by four G₄ repeats with three T₄ connectors has been shown by NMR analysis (Wang & Patel, 1995) to form an intramolecular quadruplex with two lateral T₄ and one diagonal T₄ loop, closely resembling the topology of the intramolecular quadruplex formed by four TTAGGG human telomeric repeats (Wang & Patel, 1993: this structure is discussed in detail in Chapter 3).

Several studies have examined the effects of systematic changes of sequence on structure using the d(G₄T₄G₄) quadruplex as a starting point (Table 2–1). The d(G₃T₄G₃) bimolecular quadruplex (Smith, Lau & Feigon, 1994; Keniry *et al.*, 1995) has, as expected, the same overall topology as d(G₄T₄G₄), with three stacked G-quartets. However, in d(G₃T₄G₃) the glycosidic angles in the strands are asymmetric, so that the resultant stacking is also asymmetric: one terminal G-quartet has all guanosines having a *syn* conformation while the other two both have two *syn* and two *anti*. Changing the central T₄ loop to a T₃ one in d(G₄T₃G₄) (and also in the bromine-containing sequence d(G₄U^{Br}T₂G₄) used to phase the crystal structure) results in bimolecular quadruplexes with two lateral loops (Hazel,

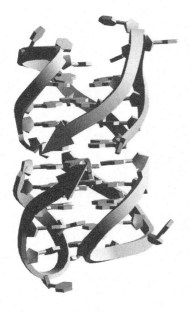

FIGURE 2–11 The head-to-head bimolecular quadruplex observed in the crystal structure of the d[$G_4T_3G_4$] quadruplex (Hazel, Parkinson & Neidle, 2006). The arrangement in the crystal is shown here, with two quadruplexes forming a dimer.

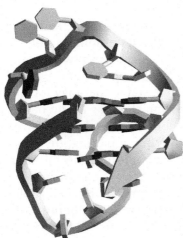

FIGURE 2–12 Cartoon representation of one of the ensemble of NMR structures of d($G_3T_4G_4$) (Šket, Črnugelj & Plavec, 2003).

Parkinson & Neidle, 2006). Several variants were observed in the crystal structures, with both head-to-head and head-to-tail stacked dimers (Figures 2–5c, 2–11). The marked preference for lateral rather than diagonal loops is probably a consequence of the conformational limitations of three nucleotides in a loop. More profound topological changes occur when the number of guanines is diminished, as found in the NMR structures of d($G_3T_4G_4$) (Črnugelj, Šket & Plavec, 2003) and d($G_4T_4G_3$) (Črnugelj, Hud & Plavec, 2002). These structures are relevant to the wider question of topology in genomic sequences, where unequal numbers of guanines in successive G-tracts do sometimes occur—see

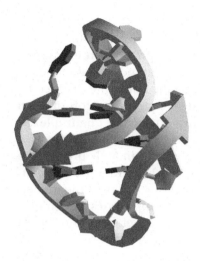

FIGURE 2–13 Cartoon representation of one of the ensemble of NMR structures of d(G₄T₄G₃) (Čmugelj, Hud & Plavec, 2002).

Chapters 1 and 6). Both structures are asymmetric, which is unsurprising given the unequal number of guanines on each side of the sequences. There are three stacked G-quartets in the d(G$_3$T$_4$G$_4$) quadruplex (Figure 2–12) with a necessarily asymmetric pattern of glycosidic angles, one diagonal and one lateral loop, which overall produces an unusual pattern of three parallel and one anti-parallel strand orientations, which is independent of the nature of the cation. The "isomeric" quadruplex formed by d(G$_4$T$_4$G$_3$) has, on the other hand, two asymmetric diagonal loops (Figure 2–13), which appear to be favored in the presence of Na$^+$ ions. Both structures have excess stacked guanine bases that are not part of the G-quartets, but which could conceivably form part of ligand binding sites in quadruplexes with these characteristics.

Quadruplex Structure Prediction

The prediction of quadruplex topology and stability from knowledge of sequence alone is ultimately key to understanding the large amount of bioinformatics data on quadruplex sequence occurrence (Chapter 1). Experimental structural data provide an essential starting point, although there is as yet very little data on the modeling of unknown structures based on existing ones, analogous to protein homology modeling. It is not known *a priori* whether one nucleotide can be exchanged for another in a given quadruplex structure, and indeed the limited data (for example, on *Oxytricha*-like quadruplexes discussed above) suggest that even conservative sequence changes can result in major topological differences. There have been very few attempts at quadruplex structure prediction, at least disclosed publicly. Two studies (Stegle *et al.*, 2009; Wong *et al.*, 2010) have used a regression analysis approach to attempt to predict relative stabilities of all quadruplexes in the human genome, based on the earlier bioinformatics analyses

(Huppert & Balasubramanian, 2005; Todd, Johnston & Neidle, 2005) that found over 350,000 occurrences of sequences based on the quadruplex sequence definition given earlier in this chapter. While this statistical approach shows promise when examining large numbers of quadruplexes, its relevance to the prediction of an individual sequence with an as yet unknown fold remains to be shown. An interesting and quite distinct yet practical approach (Fogolari *et al.*, 2009) has used the available structural information from the PDB to construct structures for as yet unknown intramolecular quadruplexes, also based on the "standard" sequence using molecular modeling techniques to assemble the known fragments together. This method will undoubtedly ultimately prove to be powerful once many more structures are determined, since only 14 intramolecular quadruplex structures were available for this study. Even so a total of 4,418 possible models were generated from these data. These are dominated by parallel topologies, which may reflect the number in the experimental database, or it may be a real effect; a number of topologies were ruled out as a consequence of steric clashes. Overall the approach when tested against a quadruplex example sequence of known topology appears to give useful predictions, even with the small learning set. However, it is clear that the approach is heavily dependent on quadruplex folding adopting the simple rules outlined in this chapter. The ability to predict added complexities such as the occurrence of the insertion of guanines in the loops into the core G-quartets (observed in some genomic quadruplexes—see Chapter 6) will require much greater knowledge of quadruplex structure and behavior, and much more experimental biophysical and structural data than are currently available.

Quadruplex Loops

The available database of quadruplex structures, taken with data on the preferences for connecting sequences in the human genome (Todd, Johnston & Neidle, 2005), indicate that these loop sequences play a key role in defining quadruplex topology. Much effort has gone into understanding the factors involved. It has been suggested (Webba da Silva, 2007; Webba da Silva et al. 2009; Fogolari *et al.*, 2009) that some potential loop combinations are sterically implausible, and that just 13 possible loop combinations are likely to be stable for three-loop intramolecular quadruplexes having clockwise polarity for the strand direction. There are another 13 possibilities that complement this set, which have the opposite strand polarity. It has also been proposed that glycosidic angle, groove width, the nature of a particular loop, and overall topology are interdependent, although firm correlations must await more extensive high-resolution data on a greater variety of quadruplex topologies than are currently known. It remains the case, however, that only a small number of the possibilities for intramolecular quadruplex folds have been observed to date, suggesting that these represent at least some of the most stable arrangements. Others have not been observed, probably in part because of unacceptable steric clashes between loop residues that result in reduced overall stability. Rather more likely is that novel stable topologies will emerge that involve insertion of guanine-containing connector residues into the G-quartet cores, once the data set becomes substantially larger.

It is now possible to define a set of rules that specifies the influence of short loops on the overall topology of G-quadruplexes. There are several significant contributing factors, notably loop length and sequence, as well as G-tract length. In general bimolecular quadruplex topology appears not to be markedly dependent on the nature of the cation, in striking contrast with unimolecular quadruplexes—examples are the *Oxytricha* bimolecular quadruplex compared with the human telomeric intramolecular quadruplexes (see Chapter 3 for more detail). In general the shorter a loop, the more likely it is to be a propeller form (Hazel *et al.*, 2004), which is explicable in purely steric terms since a diagonal or lateral loop comprising a single nucleotide is too short to either span three G-quartets (diagonal) or the length of a G-quartet (lateral). Propeller single-nucleotide loops have a sequence preference that has been observed at the genomic level (Todd, Johnston & Neidle, 2005), for a pyrimidine over a purine nucleotide. This has also been experimentally observed (Guédin *et al.*, 2008). Single-nucleotide loops confer enhanced stability on quadruplexes (Hazel *et al.*, 2004; Risitano & Fox, 2004; Bugaut & Balasubramanian, 2008; Smargiasso *et al.*, 2008) and the presence of one or two then results in a strong preference for the parallel topology.

The behavior of quadruplexes with longer loops continues to be more challenging to understand and predict. It has been widely assumed, on the basis of a number of studies (for example, Hazel et al, 2004; Bugaut & Balasubramanian, 2008), that loop lengths have a maximum length of seven nucleotides, and indeed all the bioinformatics studies on genomic quadruplex occurrence are predicated on this being an upper limit. Results from a careful biophysical study (Guédin, Alberti & Mergny, 2010) suggest that this is not always the case and that in certain circumstances, notably in the presence of two very short loops, a third can be very long (>10 nucleotides). Presumably this is due to stabilization of loop secondary structures.

There are also sequence preferences within mid-size loops as indicated by both genomic surveys (Todd, Johnston & Neidle, 2005; Todd & Neidle, 2011), and experimental observations (Guédin, Alberti & Mergny, 2009). The latter approach has found that adenosines are strongly disfavored when at the first position in the second loop within a quadruplex containing all three-nucleotide loops. Whether this is the case in all quadruplex types remains to be fully determined. Molecular dynamics simulations have been employed to model the stability of particular loop arrangements in quadruplexes and their features have been recognized as presenting particular challenges for simulation approaches (Fadrná *et al.*, 2009), especially for propeller loops. Thus the experimentally crystallographically determined conformations of three-nucleotide propeller loops are not seen even with long time-scale simulations. This is not unsurprising since although the crystallographic observations of native TTA loops (Parkinson, Lee & Neidle, 2002) indicate that their conformations are conserved, other crystal structures involving ligand complexes demonstrate that they are highly flexible, with the inference that there are multiple local minima in close energetic proximity. A further theoretical analysis (Cang,

Šponer & Cheatham, 2011) of anti-parallel quadruplexes with varying lengths of G-tracts has highlighted their effects on patterns of glycosidic angles. In particular G-tracts with two or four guanines tend to produce quadruplexes with greater conformational stability than those with three guanines. This is consistent with the experimental observations of conformational conservation for the *Oxytricha* d($G_4T_4G_4$) quadruplex, independent of the nature of the cation, whereas human telomeric quadruplexes (which contain G3 tracts) appear to be more prone to cation-induced polymorphism, at least in relatively dilute solution. The overwhelming majority of experimental studies on quadruplexes in solution have been performed in dilute (<5 mM) conditions. These are not necessarily the best indicators of behavior in cellular environments, as will be discussed in Chapter 3.

General Reading

Burge, S., Parkinson, G.N., Hazel, P., Todd, A.K., Neidle, S., 2006. Quadruplex DNA: sequence, topology and structure. Nucleic Acids Res. 34, 5402–5415.

Davies, J.T., 2004. G-tetrads 40 years later: from 5'-GMP to molecular biology and supramolecular chemistry. Angew. Chem. Intl. Edit. 43, 668–698.

Gilbert, D.E., Feigon, J., 1999. Multistranded DNA structures. Curr. Opin. Struct. Biol. 9, 305–314.

Neidle, S., Balasubramanian, S. (Eds.), 2006. Quadruplex Nucleic Acids. Royal Society of Chemistry, Cambridge, UK.

Neidle, S., Parkinson, G.N., 2008. Quadruplex DNA crystal structures and drug design. Biochimie 90, 1184–1196.

Neidle, S., 2009. The structures of quadruplex nucleic acids and their drug complexes. Curr. Opin. Struct. Biol. 19, 1–12.

Simonsson, T., 2001. G-quadruplex DNA structures—variations on a theme. Biol. Chem. 382, 621–628.

Williamson, J.R., 1994. G-quartet structures in telomeric DNA. Ann. Rev. Biophy. 23, 703–730.

References

Aboul-ela, F., Murchie, A.I.H., Norman, D.G., Lilley, D.M.J., 1994. Solution structure of a parallel-stranded tetraplex formed by d(TG$_4$T) in the presence of sodium ions by nuclear magnetic resonance spectroscopy. J. Mol. Biol. 243, 438–471.

Arnott, S., Chandrasekaran, R., Marttila, C.M., 1974. Structures for polyinosinic acid and polyguanylic acid. Biochem. J. 141, 537–543.

Balagurumoorthy, P., Brahmachari, S.K., 1994. Structure and stability of human telomeric sequence. J. Biol. Chem. 269, 21858–21869.

Bugaut, A., Balasubramanian, S., 2008. A sequence-independent study of the influence of short loop lengths on the stability and topology of intramolecular DNA quadruplexes. Biochemistry 47, 689–697.

Cáceres, C., Wright, G., Gouyette, C., Parkinson, G., Subirana, J.A., 2004. A thymine tetrad in d(TGGGGT) quadruplexes stabilized with Tl$^+$/Na$^+$ ions. Nucleic Acids Res. 32, 1097–1102.

Campbell, N.H., Smith, D.L., Reszka, A.P., Neidle, S., O'Hagan, D., 2011. Fluorine in medicinal chemistry: β-fluorination of peripheral pyrrolidines attached to acridine ligands affects their interactions with G-quadruplex DNA. Org. Biomol. Chem. 9, 1328–1331.

Cang, X., Šponer, J., Cheatham III, T.E., 2011. Explaining the varied glycosidic conformational, G-tract length and sequence preferences for anti-parallel G-quadruplexes. Nucleic Acids Res. In press

Chen, F.M., 1992. Sr^{2+} facilitates intermolecular G-quadruplex formation of telomeric sequences. Biochemistry 31, 3769–3776.

Črnugelj, M., Hud, N.V., Plavec, J., 2002. The solution structure of $d(G_4T_4G_3)_2$: a bimolecular G-quadruplex with a novel fold. J. Mol. Biol. 320, 911–924.

Črnugelj, M., Šket, P., Plavec, J., 2003. Small change in a G-rich sequence, a dramatic change in topology: new dimeric G-quadruplex folding motif with unique loop orientations. J. Amer. Chem. Soc. 125, 7866–7871.

Creze, C., Rinaldi, B., Haser, R., Bouvet, P., Gouet, P., 2007. Structure of a d(TGGGGT) quadruplex crystallized in the presence of Li^+ ions. Acta Crystallogr. D63, 682–688.

Espositoa, V., Galeonea, A., Mayola, L., Olivieroa, G., Virgilioa, A., Randazzo, L., 2007. A topological classification of G-quadruplex structures. Nucleos. Nucleot. Nucleic Acids 26, 1155–1159.

Fadrná, E., Špačková, N., Sarzyňska, J., Koča, J., Orozco, M., Cheatham III, T.E., et al., 2009. Single stranded loops of quadruplex DNA as key benchmark for testing nucleic acids force field. J. Chem. Theory Comp. 5, 2514–2530.

Fogolari, F., Haridas, H., Corazza, A., Viglino, P., Corã, D., Caselle, M., et al., 2009. Molecular models for intrastrand DNA G-quadruplexes. BMC Struct. Bio. 9, 64.

Gellert, M., Lipsett, M.N., Davies, D.R., 1962. Helix formation by guanylic acid. Proc. Natl. Acad. Sci. USA 48, 2013–2018.

Gill, M.L., Strobel, S.A., Loria, J.P., 2005. [205]Tl NMR methods for the characterization of monovalent cation binding to nucleic acids. J. Amer. Chem. Soc. 127, 16723–16732.

Gill, M.L., Strobel, S.A., Loria, J.P., 2006. Crystallization and characterization of the thallium form of the *Oxytricha* nova G-quadruplex. Nucleic Acids Res. 34, 4506–4514.

Guédin, A., Alberti, P., Mergny, J.-L., 2009. Stability of intramolecular quadruplexes: sequence effects in the central loop. Nucleic Acids Res. 37, 5559–5567.

Guédin, A., Alberti, P., Mergny, J.-L., 2010. How long is too long? Effects of loop size on G-quadruplex stability. Nucleic Acids Res. 38, 7858–7868.

Guédin, A., De Cian, A., Gros, J., Lacroix, L., Mergny, J.-L., 2008. Sequence effects in single-base loops for quadruplexes. Biochimie 90, 686–696.

Haider, S., Parkinson, G.N., Neidle, S., 2002. Crystal structure of the potassium form of an *Oxytricha nova* G-quadruplex. J. Mol. Biol. 320, 189–200.

Hardin, C.C., Henderson, E., Watson, T., Prosser, J.K., 1991. Monovalent cation induced structural transitions in telomeric DNAs: G-DNA folding intermediates. Biochemistry 30, 4460–4472.

Hazel, P., Huppert, J., Balasubramanian, S., Neidle, S., 2004. Loop-length-dependent folding of G-quadruplexes. J. Amer. Chem. Soc. 126, 16405–16415.

Hazel, P., Parkinson, G.N., Neidle, S., 2006. Topology variation and loop structural homology in crystal and simulated structures of a bimolecular DNA quadruplex. J. Amer. Chem. Soc. 128, 5480–5487.

Horvath, M.P., Schultz, S.C., 2001. DNA G-quartets in a 1.86 Å resolution structure of an *Oxytricha nova* telomeric protein-DNA complex. J. Mol. Biol. 310, 367–377.

Howard, F.B., Miles, H.T., 1982. Poly(inosinic acid) helices: essential chelation of alkali metal ions in the axial channel. Biochemistry 21, 6736–6745.

Howard, F.B., Frazier, J., Miles, H.T., 1977. Stable and metastable forms of poly(G). Biopolymers 16, 791–809.

Hud, N.V., Plavec, J., 2006. The role of cations in determining quadruplex structures and stability. In: Neidle, S., Balasubramanian, S. (Eds.), Quadruplex Nucleic Acids (pp. 100–130). RSC Publishing, Cambridge UK.

Hud, N.V., Schultze, P., Feigon, J., 1998. Ammonium ion as an NMR probe for monovalent cation coordination sites of DNA quadruplexes. J. Amer. Chem. Soc. 120, 6403–6404.

Huppert, J.L., Balasubramanian, S., 2005. Prevalence of quadruplexes in the human genome. Nucleic Acids Res. 35, 406–413.

Ida, R., Wu, G., 2008. Direct NMR detection of alkali metal ions bound to G-quadruplex DNA. J. Amer. Chem. Soc. 130, 3590–3602.

Ida, R., Kwan, I.C., Wu, G., 2007. Direct ^{23}Na NMR observation of mixed cations residing inside a G-quadruplex channel. Chem. Commun. (Cam.), 795–797.

Kang, C., Xhang, X., Ratliff, R., Moysis, R., Rich, A., 1992. Crystal structure of *Oxytricha* telomeric DNA. Nature 356, 126–131.

Keniry, M.A., Strahan, G.D., Owen, E.A., Shafer, R.H., 1995. Solution structure of the Na$^+$ form of the dimeric guanine quadruplex [d(G$_3$T$_4$G$_3$)]$_2$. Eur. J. Biochem. 233, 631–643.

Kwan, I.C., She, Y.M., Wu, G., 2007. Trivalent lanthanide metal ions promote formation of stacking G-quartets. Chem. Commun., 4286–4288.

Laughlin, G., Murchie, A.I.H., Norman, D.G., Moore, M.H., Moody, P.C.E., Lilley, D.M.J., et al., 1994. The high-resolution crystal structure of a parallel-stranded guanine tetraplex. Science 265, 520–524.

Lee, M.P., Parkinson, G.N., Hazel, P., Neidle, S., 2007. Observation of the coexistence of sodium and calcium ions in a DNA G-quadruplex ion channel. J. Amer. Chem. Soc. 129, 10106–10107.

Marincola, F.C., Virno, A., Randazzo, A., Mocci, F., Saba, G., Lai, A., 2009. Competitive binding exchange between alkali metal ions (K$^+$, Rb$^+$, and Cs$^+$) and Na$^+$ ions bound to the dimeric quadruplex [d(G$_4$T$_4$G$_4$)]$_2$: a ^{23}Na and ^{1}H NMR study. Magn. Reson. Chem. 47, 1036–1042.

Meyer, M., Brandl, M., Sühnel, J., 2001. Are guanine tetrads stabilized by bifurcated hydrogen bonds? J. Phys. Chem. A 105, 8223–8225.

Oka, Y., Thomas Jr., C.A., 1987. The cohering telomeres of Oxytricha. Nucleic Acids Res. 15, 8877–8898.

Parkinson, G.N., Lee, M.P.H., Neidle, S., 2002. Crystal structure of parallel quadruplexes from human telomeric DNA. Nature 417, 876–880.

Pedroso, I.M., Duarte, L.F., Yanez, G., Baker, A.M., Fletcher, T.M., 2007. Induction of parallel human telomeric G-quadruplex structures by Sr$^{(2+)}$. Biochem. Biophys. Res. Commun. 358, 298–303.

Pérez, A., Marchán, I., Svozil, D., Šponer, J., Cheatham 3rd, T.E., Laughton, C.A., et al., 2007. Refinement of the AMBER force field for nucleic acids: improving the description of alpha/gamma conformers. Biophys. J. 92, 3817–3829.

Phan, A.T., 2010. Human telomeric G-quadruplex: structures of DNA and RNA sequences. FEBS J. 277, 1107–1117.

Phan, A.T., Kuryavyi, V., Patel, D.J., 2006. DNA architecture: from G to Z. Curr. Opin. Struct. Biol. 16, 1–11.

Phillips, K., Dauter, Z., Murchie, A.I., Lilley, D.M., Luisi, B., 1997. The crystal structure of a parallel-stranded guanine tetraplex at 0.95 Å resolution. J. Mol. Biol. 273, 171–182.

Qin, Y., Hurley, L.H., 2008. Structures, folding patterns and functions of intramolecular DNA G-quadruplexes found in eukaryotic promoter regions. Biochimie 90, 1149–1171.

Risitano, A., Fox, K.R., 2004. Influence of loop size on the stability of intramolecular DNA quadruplexes. Nucleic Acids Res. 32, 2598–2606.

Schultze, P., Hud, N.V., Smith, F.W., Feigon, J., 1999. The effect of sodium, potassium and ammonium ions on the conformation of the dimeric quadruplex formed by the *Oxytricha* nova telomere repeat oligonucleotide d(G$_4$T$_4$G$_4$). Nucleic Acids Res. 27, 3018–3028.

Schultze, P., Smith, F.W., Feigon, J., 1994. Refined solution structure of the dimeric quadruplex formed from the *Oxytricha* telomeric oligonucleotide d(GGGGTTTTGGGG). Structure 2, 221–233.

Sen, D., Gilbert, W., 1988. Formation of parallel four-stranded complexes by guanine-rich motifs in DNA and its implications for meiosis. Nature 334, 364–366.

Sen, D., Gilbert, W., 1990. A sodium-potassium switch in the formation of four-stranded G4-DNA. Nature 344, 410–414.

Sket, P., Plavec, J., 2010. Tetramolecular DNA quadruplexes in solution: insights into structural diversity and cation movement. J. Amer. Chem. Soc. 132, 12724–12732.

Smargiasso, N., Rosu, F., Hsia, W., Closon, P., Baker, E.S., Bowers, M.T., et al., 2008. G-quadruplex DNA assemblies: loop length, cation identity and multimer formation. J. Amer. Chem. Soc. 130, 10208–10216.

Smirnov, I.V., Shafer, R.H., 2000. Lead is unusually effective in sequence-specific folding of DNA. J. Mol. Biol. 296, 1–5.

Smirnov, I.V., Kotch, F.W., Pickering, I.J., Davis, J.T., Shafer, R.H., 2002. Pb EXAFS studies on DNA quadruplexes: identification of metal ion binding site. Biochemistry 41, 12133–12139.

Smith, F.W., Feigon, J., 1992. Quadruplex structure of *Oxytricha* telomeric DNA oligonucleotides. Nature 356, 164–168.

Smith, F.W., Lau, F.W., Feigon, J., 1994. d($G_3T_4G_3$) forms an asymmetric diagonally looped dimeric quadruplex with guanosine 5′-*syn-syn-anti* and 5′-*syn*-anti-*anti* N-glycosidic conformations. Proc. Natl. Acad. Sci. USA 91, 10546–10550.

Špačková, N., Berger, I., Šponer, J., 1999. Nanosecond molecular dynamics simulations of parallel and antiparallel guanine quadruplex DNA molecules. J. Amer. Chem. Soc. 121, 5519–5534.

Špačková, N., Berger, I., Šponer, J., 2001. Structural dynamics and cation interactions of DNA quadruplex molecules containing mixed guanine/cytosine quartets revealed by large-scale MD simulations. J. Amer. Chem. Soc. 123, 3295–3307.

Šponer, J., Špačková, N., 2007. Molecular dynamics simulations and their application to four-stranded DNA. Methods 43, 278–290.

Stegle, O., Payet, L., Mergny, J.-L., MacKay, D.J.C., Huppert, J.L., 2009. Predicting and understanding the stability of G-quadruplexes. Bioinformatics 25, 1374–1382.

Sundquist, W.I., Klug, A., 1989. Telomeric DNA dimerizes by formation of guanine tetrads between hairpin loops. Nature 342, 825–829.

Todd, A.K., Neidle, S., 2011. Mapping the sequences of potential guanine quadruplex motifs. Nucleic Acids Res. DOI: 10.1093/nar/gkr104.

Todd, A.K., Johnston, M., Neidle, S., 2005. Highly prevalent putative guanine quadruplex sequence motifs in human DNA. Nucleic Acids Res. 33, 2901–2907.

van Mourik, T., Dingley, A.J., 2005. Characterization of the monovalent ion position and hydrogen-bond network in guanine quartets by DFT calculations of NMR parameters. Chemistry 11, 6064–6079.

Walmsley, J.A., Burnett, J.F., 1999. A new model for the K+-induced macromolecular structure of guanosine 5′-monophosphate in solution. Biochemistry 38, 14063–14068.

Wang, Y., Patel, D.J., 1995. Solution structure of the *Oxytricha* telomeric repeat d[$G_4(T_4G_4)_3$] G-tetraplex. J. Mol. Biol. 251, 76–94.

Wang, Y., Patel, D.J., 1993. Solution structure of the human telomeric repeat d[$AG_3(T_2AG_3)_3$] G-tetraplex. Structure 1, 263–282.

Webba da Silva, M., 2007. Geometric formalism for DNA quadruplex folding. Chemistry 13, 9738–9745.

Webba da Silva, M., Trajkovski, M., Sannohe, Y., Ma'ani Hessari, N., Sugiyama, H., Plavec, J., 2009. Design of a G-quadruplex topology through glycosidic angles. Angew. Chem. Int. Edit. 48, 9167–9170.

Williamson, J.R., Raghuraman, M.K., Cech, T.R., 1989. Monovalent cation-induced structure of telomeric DNA: the G-quartet model. Cell 45, 167–176.

Wong, H.M., Stegle, O., Rodgers, S., Huppert, J.L., 2010. A toolbox for predicting G-quadruplex formation and stability. J. Nucleic Acids, 564946.

Wu, G., Kwan, I.C., 2009. Helical structure of disodium 5′-guanosine monophosphate self-assembly in neutral solution. J. Amer. Chem. Soc. 131, 3180–3182.

Zimmerman, S.B., 1976. X-ray study by fiber diffraction methods of a self-aggregate of guanosine-5′-phosphate with the same helical parameters as poly(rG). J. Mol. Biol. 106, 663–672.

Zimmerman, S.B., Cohen, G.H., Davies, D.R., 1975. X-ray fiber diffraction and model-building study of polyguanylic acid and polyinosinic acid. J. Mol. Biol. 92, 181–192.

3

The Structures of Human Telomeric DNA Quadruplexes

Introduction

The concept of human telomeric quadruplexes as therapeutic targets is described in detail in Chapters 1 and 6. Structure-based drug design, or studies aiming to relate activity of a series of compounds to structure, require as a starting point structural data on the target. As yet there is no experimental structural information that directly reflects telomeres in their biological environment, either on the telomeric D-loop, which may involve a quadruplex-type junction, or more pertinently on a substantial length of the single-stranded telomeric DNA overhang (although crystal structures are available for the hPOT1 protein bound to the single-stranded oligonucleotide sequence d(TTAGGGTTAG) (Lai, Podell & Cech, 2004), and more recently with the sequence d(GGTTAGGGTTAG) (Nandamukar, Cech & Podell, 2010)). These structures show the overhang in a natural single-stranded conformation (Figure 3–1). In the absence of the POT1 protein (for example, when displaced by a quadruplex-binding small molecule) the single-stranded DNA will have a natural tendency to fold into quadruplex structures, especially since the nucleus has an intracellular ionic environment that would provide a sufficient K^+ ionic concentration (~150 mM) to stabilize quadruplex structures. The single-stranded overhang is also a stretch of DNA sequence in the genome that is unique in being free from the structural constraints of the double helix.

An overhang length of, say, 120 nucleotides, comprising 20 telomeric repeats, can in principle form 3–5 quadruplex structures. However, as yet we do not have any detailed structural information on how such an array is organized, although there have been several molecular modeling studies that demonstrate some possibilities. Instead there are a number of crystal and NMR studies of single telomeric quadruplexes, all comprising four telomeric repeats, albeit with differing terminal nucleotides. This added feature, of variability at the termini, has been used either to ensure crystallization or interpretable single-species NMR spectra. These structures (Table 3–1), discussed below, are all relevant models for quadruplexes formed from the single-stranded overhang. Although they do display a number of important features that reflect features in the natural telomeric structures, equally others, particularly involving multimeric quadruplex arrangements, can currently be only modeled or inferred. It is with that caveat that we discuss the current state of knowledge of telomeric quadruplexes.

Therapeutic Applications of Quadruplex Nucleic Acids. DOI: 10.1016/B978-0-12-375138-6.00003-0

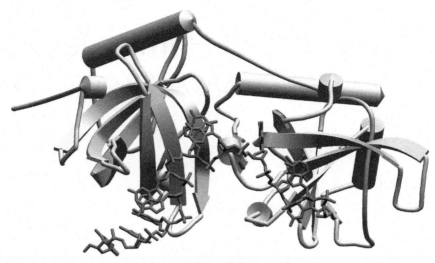

FIGURE 3–1 Representation of the crystal structure (PDB id 3KJP) of the protein–DNA complex involving the POT1 protein and the single-stranded sequence d(GGTTAGGGTTAG) (Nandamukar, Cech & Podell, 2010). The protein is shown in cartoon form and the single-stranded DNA as shaded sticks. The DNA has a number of regions of local structure, with base–base and base–phenylalanine stacking at several positions.

The hexanucleotide human telomeric DNA repeat of d(TTAGGG) can be the building block for all three categories of quadruplex type, unimolecular, bimolecular, and tetramolecular (Patel, Phan & Kuryavyi, 2007). The greatest attention has focused on unimolecular species since these are considered to be the category most likely to be formed by folding of the 3′-single-stranded overhang of telomeric DNA. Unimolecular quadruplexes can also be formed from duplex telomeric DNA when it is transiently single-stranded during replication or transcription—quadruplex formation has been indirectly observed by gel electrophoresis methods in G-tract-containing duplex DNA sequences during *in vitro* transcription (Zheng *et al.*, 2010). G-quadruplex formation has been reported as condensed objects formed, with G-quadruplex dimensions, along G-rich DNA and DNA/RNA heteroduplexes using electron microscopy (Duquette *et al.*, 2004) and by atomic force microscopy (AFM) (Neaves *et al.*, 2009), and was shown to be an effective hindrance to transcriptional activity. However, bimolecular quadruplexes are also of biological relevance since the extreme two telomeric end-repeats of the single-stranded overhang can readily associate with a second overhang molecule (Figure 3–2), and the resulting structure may serve as a model for chromosomal fusions and anaphase bridges. Both monomolecular and bimolecular species can also be conceivably formed by extrusion from a duplex telomeric DNA sequence (see, for example, Risitano & Fox, 2003; Zhou *et al.*, 2009; Tran, Mergny & Alberti, 2011). The equilibrium in duplex telomeric DNA does not naturally favor quadruplex formation since the complementary

Table 3–1   Human telomeric DNA quadruplex structures available in the Protein Data Bank. Loop types are designated as L (lateral), D (diagonal) and P (parallel). The nucleoside I in the penultimate entry represents inosine

Sequence	Method	Topology	Loops	PDB id	References
d[AG$_3$(T$_2$AG$_3$)$_3$]	NMR	anti-parallel	1 × D, 2 × L	143D	Wang & Patel, 1992
d[AG$_3$(T$_2$AG$_3$)$_3$]	X-ray	parallel	3 × P	1KF1	Parkinson, Lee & Neidle, 2002
d[UAG$_3$UTAG$_3$T]	X-ray	parallel	2 × P	1K8P	Parkinson, Lee & Neidle, 2002
d[TAG$_3$(T$_2$AG$_3$)$_3$]	NMR	parallel	3 × P	XXX	Heddi & Phan, 2011
d[G$_2$T$_2$AG$_3$T$_2$AG$_3$T + TAG$_3$U]	NMR	hybrid-1 (3 + 1)	1 × P, 2 × L	2AQY	Zhang, Phan & Patel, 2005
d[T$_2$G$_3$(T$_2$AG$_3$)$_3$A]	NMR	hybrid-1 (3 + 1)	1 × P, 2 × L	2KGU	Luu *et al.*, 2006
d[A$_3$G$_3$(T$_2$AG$_3$)$_3$AA]	NMR	hybrid-1 (3 + 1)	1 × P, 2 × L	2HY9	Dai *et al.*, 2007b
d[TAG$_3$(T$_2$AG$_3$)$_3$]	NMR	hybrid-1 (3 + 1)	1 × P, 2 × L	2JSM	Phan *et al.*, 2007
d[(T$_2$AG$_3$)$_4$TT]	NMR	hybrid-2 (3 + 1)	2 × L, 1 × P	2JPZ	Dai *et al.*, 2007a
d[TAG$_3$(T$_2$AG$_3$)$_3$TT]	NMR	hybrid-2 (3 + 1)	2 × L, 1 × P	2JSL	Phan *et al.*, 2007
d[(GGGTTA)$_3$GGGT]	NMR	basket	1 × D, 2 × L	2KF7	Lim *et al.*, 2009b
d[A(G$_3$TTA)$_2$IG$_2$TTAG$_3$T]	NMR	basket	1 × D, 2 × L	2KKA	Zhang *et al.*, 2010
d[A(G$_3$CTA)$_3$G$_3$]	NMR	anti-parallel	3 × L	2KM3	Lim *et al.*, 2009a

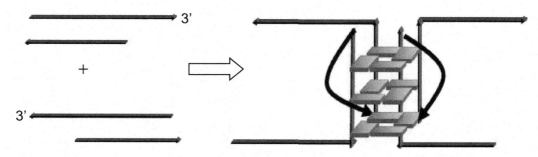

FIGURE 3–2 Schematic view of a bimolecular quadruplex formed at the interface between two separate telomeric DNA strands, with each half-quadruplex formed by folding of the single-stranded overhang.

strand would have to form a C-rich i-motif arrangement, which is only favored at low pH (Phan & Mergny, 2002; Alberti & Mergny, 2003), although the equilibrium at physiological pH can also be disturbed in favor of quadruplex stabilization by small molecules that have selectivity for quadruplex over duplex.

An Intramolecular Quadruplex in a Sodium Ion Environment

The determination of the structure and topology of the 22-mer sequence d[AGGG-(TTAGGG)$_3$] in 0.1 M NaCl solution (at a 7 mM oligonucleotide concentration) by NMR and molecular dynamics modeling methods (Wang & Patel, 1993) was a landmark in our understanding of quadruplex folding. The 1-D proton NMR spectrum showed 12 well-resolved peaks in the imino region, indicating that this sequence forms a stable single species under these solution conditions. At that time the NMR and crystal structures of the *Oxytricha nova* bimolecular quadruplex appeared to have distinct topologies (see Chapter 2), suggesting that quadruplexes have an inherent tendency for polymorphism, although subsequent crystallographic studies have shown that this *Oxytricha* quadruplex does actually have a conserved structure in both the crystal and solution (see Chapter 2). The analysis of the 2-D NMR data for the human telomeric 22-mer has revealed a basket-type quadruplex structure held together by a combination of diagonal and lateral loops (Figure 3–3), with the backbone tracing a single continuous trajectory. The six refined structures constituting the ensemble are shown in Figure 3–4 (PDB id 143D). The anti-parallel fold has a core of three dynamically stable stacked G-quartets that involve all the 12 guanines in the sequence. The overall fold is conserved in all the structures of the ensemble, although some of the loop bases adopt a diversity of conformations, especially for the middle thymine nucleotide in the two lateral loops. The adenine and terminal thymine bases from each of these loops form two flexible A...T base associations that are

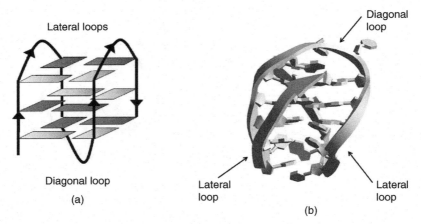

FIGURE 3–3 View of the human telomeric 22-mer NMR structure (Wang & Patel, 1993), determined in sodium ion solution. **(a)** Schematic showing the overall topology, with the individual rectangles indicating the glycosidic conformation at each guanosine nucleoside; light shading indicates an *anti* conformation and darker shading a *syn* one. **(b)** Cartoon drawn from the coordinates of one of the ensemble of six deposited NMR structures for the 22-mer, with the three TTA loops being labeled.

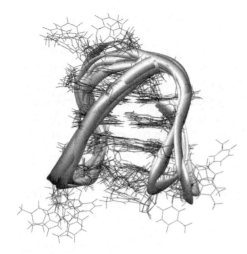

FIGURE 3–4 The complete ensemble of all six deposited structures for the 22-mer Na$^+$ NMR structure, superimposed on one another. The backbones have been drawn in cartoon form, emphasizing the same overall topology, but with some minor divergence between some structures. The very clear differences in orientation for individual loop bases indicate the flexibility of these regions.

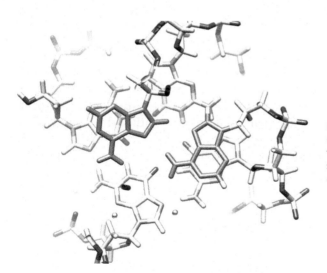

FIGURE 3–5 View of the adenines A1 and A13 (shown with shaded bonds) that form part of the two lateral loops in the 22-mer Na$^+$ NMR structure, highlighting the stacking of these adenines onto the terminal G-quartet.

stacked on top of each other, and on top of the terminal G-quartet, although the adenine and thymine in each instance are not close enough to be hydrogen bonded. At the other end of the quadruplex the diagonal loop formed from residues T11, T12 and A13 is also flexible, with both thymine bases swung out from the quadruplex core and free to adopt a variety of conformations. The 5′ terminal adenine (A1) and loop A13 both stack onto the terminal G-quartet (Figure 3–5) although there is no significant hydrogen bonding between them. Presumably this stacking helps to stabilize the overall structure and is probably an important contributor to the overall conformational homogeneity of the structure. Overall the fold is close to what one would intuitively imagine being formed on the basis of a purely paper exercise, with adjacent strands all having mutually anti-parallel orientations. The guanosine glycosidic angles alternate between *anti* and *syn*, starting with the first guanosine in the sequence (G2) having an *anti* conformation. Thus each G-quartet has the terminal G-quartets with *anti-syn-anti-syn* glycosidic angles and the central one with *syn-anti-syn-anti* ones. This asymmetry in glycosidic angles results in a marked asymmetry in groove dimensions (see Chapter 9), with two medium one narrow and one wide groove (Figures 3–3b, 3–6).

Crystal Structures of Human Telomeric Quadruplexes in a Potassium Ion Environment

The crystal-structure determination (Parkinson, Lee & Neidle, 2002) of the identical 22-mer sequence d[AGGG(TTAGGG)$_3$], with crystals grown in the presence of potassium ions (50 mM) and using polyethylene glycol (PEG) 400 as a crystallizing agent, has revealed a radically distinct structure with all four oligonucleotide backbone edges of the structure in the same (i.e. parallel) orientation to each other. This is a heavily hydrated crystal with approximately 50% solvent content, although at the resolution of the structure, 2.1Å, only 68 water molecules per quadruplex were located. The overall shape of the quadruplex is of a flattened disk, contrasting with the compact globular shape of the sodium form. The three TTA loops that project out from the sides of the central three G-quartets that form the central core of the structure do so in a propeller-like manner rather than being projected outwards from the ends as are diagonal or lateral loops (Figures 3–7, 3–8). This type of loop had been observed previously in a quadruplex aptamer (Kuryavyi *et al.*, 2001).

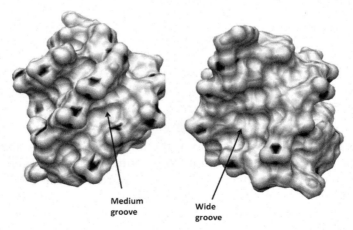

Medium groove

Wide groove

FIGURE 3–6 View of two grooves in the 22-mer Na$^+$ NMR structure, with a solvent-accessible surface shown for the structure and the surface areas corresponding to phosphorus atoms shaded in black. The variations in groove widths are clearly visible.

FIGURE 3–7 View in cartoon form, and projected onto the plane of the G-quartets, of the 22-mer crystal structure (Parkinson, Lee & Neidle, 2002). The potassium ions are shown as small spheres in the center of the G-quartet channel. The closely similar conformation of all three TTA loops is apparent.

FIGURE 3–8 Side view of the 22-mer crystal structure in cartoon representation, emphasizing its disk-like shape.

Three potassium ions were clearly visible in the electron density maps; two ions are positioned midway between pairs of G-quartet planes and the third is external to the 5′ G-quartet but is midway between it and the other 5′ G-quartet in the crystallographic dimer (Figure 3–12) so that there is a continuous line of five K$^+$ ions in the central channel. All guanosine nucleoside glycosidic angles adopt the *anti* conformation and all their sugar puckers are of the C2′-*endo* type. This structure has generated considerable controversy with its relevance to the topology of human telomeric quadruplexes in solution being the focus of much subsequent dispute, experimentation, claims and counter-claims (see, for example, Li *et al.*, 2005; Xu, Noguchi & Sugiyama, 2006). These issues will be discussed below in more detail.

The 22-mer crystal structure, at 2.1Å resolution, was solved by molecular replacement methods using the crystal structure of the analogous bimolecular quadruplex (formed from two strands of the sequence d(U^{Br}AGGGU^{Br}TAGGGT)), as a starting point. This latter structure, which was reported at the same time, was determined *ab initio* by MAD phasing methods (see Chapter 10), and shows a symmetric quadruplex sitting on a crystallographic two-fold axis, with three stacked G-quartets and two TTA propeller loops

(Figure 3–9). There are thus a total of four crystallographically independent TTA loops in these two structures and remarkably, not only do all adopt very similar overall structures, within each the adenine base swung back to stack in between the two thymine bases (Figure 3–10), but the conformation of each individual loop nucleotide is highly conserved. Figure 3–11 details this in terms of the backbone conformational angles in the TTA loops of both the 12-mer and the 22-mer. The sole differences between loop conformations are in backbone angles α (by ~150–200°) and γ (by ~100°) for the terminal thymine in each loop (Neidle & Parkinson, 2008). This conservation of loop conformation

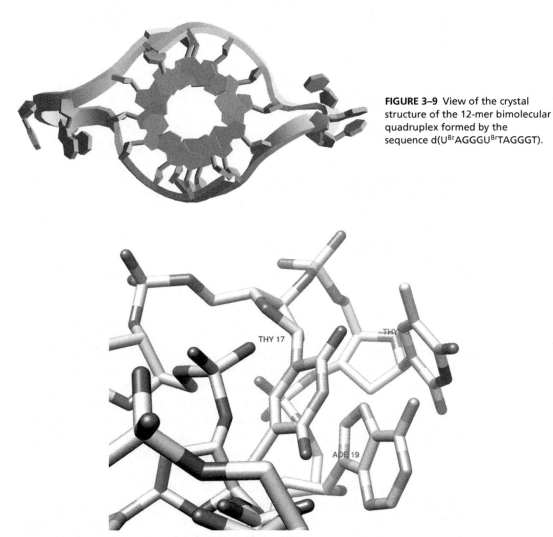

FIGURE 3–9 View of the crystal structure of the 12-mer bimolecular quadruplex formed by the sequence d(U^{Br}AGGGU^{Br}TAGGGT).

THY 17

THY

ADE 19

FIGURE 3–10 A detailed view of one TTA loop in the 22-mer crystal structure, showing the stacking between the adenine and the terminal thymine in the loop, forcing the central thymine to swing out from a stacked position.

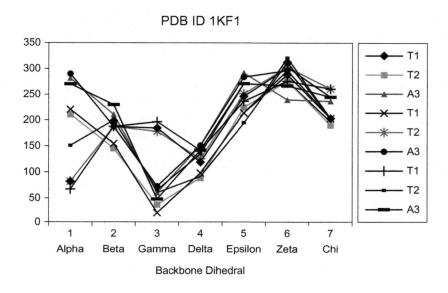

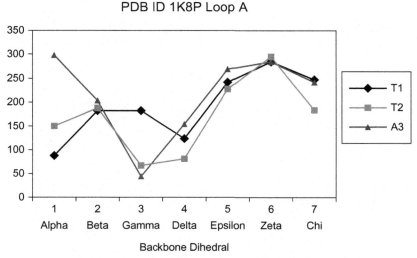

FIGURE 3–11 Plots of the loop backbone and glycosidic torsion angles for all the TTA loops in the crystal structures of the 12- and 22-mers (adapted from Neidle & Parkinson, 2008).

appears to be largely independent of crystal packing forces, even though the loops actively participate in diverse intermolecular contacts in the crystal structures of both these quadruplexes, suggesting that the crystal structures have captured a singular and stable low-energy arrangement for the loops. The crystal packing in both structures does not involve continuous columns of stacked quadruplexes. Instead, the repeating unit in each case is a 5′-5′ end-stacked dimer (Figure 3–12), a motif that has been subsequently found in the majority of other quadruplex crystal structures, as well as in solution, especially for the analogous RNA telomeric quadruplexes (Collie *et al.*, 2010).

The 22-mer quadruplex has four approximately equi-dimensional grooves, each of width ~11 Å (Figure 3–13), whose character is quite unlike those of quadruplexes with diagonal or lateral loops (see, for example, Figures 3–14a, b). Layered on top of the standard nucleic acid groove formed between the two adjacent phosphodiester backbones, as found in those other quadruplexes and in DNA duplexes and triplexes (Neidle, 2007), are the TTA loops, which lie directly in the grooves. This has the effect of making the groove surface much less regular and with a distinctive V-shape. The fact that the bases of each loop are splayed outwards from the groove means that effectively there is a hollow center to part of the groove/loop cavity, with the adenine serving as a flexible flap over this cavity (Figure 3–15). The overall dimensions of the cavity are ~10 × 11 Å.

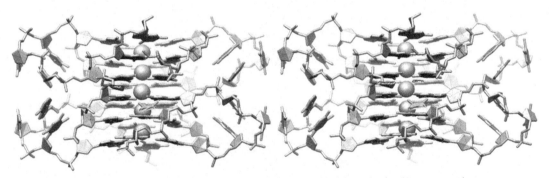

FIGURE 3–12 A view of four crystallographically related 22-mer quadruplexes in the 22-mer crystal structure (Parkinson, Lee & Neidle, 2002). The arrangement is of two head-to-head dimers, with contacts involving stacking between the central thymine bases from TTA loops. Note the uninterrupted run of five K$^+$ ions along the channel of each dimer.

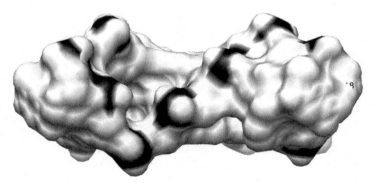

FIGURE 3–13 View into a groove along the side of the 22-mer crystal structure, drawn in solvent accessible surface representation, with the areas corresponding to phosphorus backbone atoms colored black. The groove has a V-shape, and opens up at its widest part onto the top G-quartet.

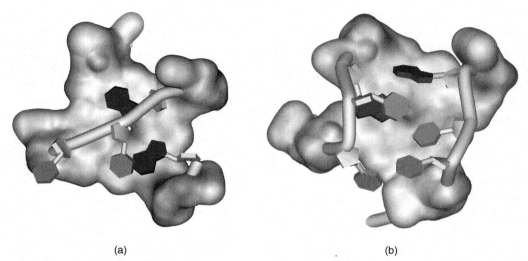

(a)   (b)

FIGURE 3–14 Views of the two G-quartet ends of the 22-mer Na$^+$ NMR structure (Wang & Patel, 1993), with the loop backbones shown in cartoon form, the G-quartet solvent-accessible surface, the loop thymine bases shown as lightly shaded hexagons and the loop adenines shown as dark shaded polygons. **(a)** shows the diagonal loop and **(b)** shows the two lateral loops. These two views suggest that small molecule binding would preferentially occur at the less hindered diagonal loop end of the quadruplex.

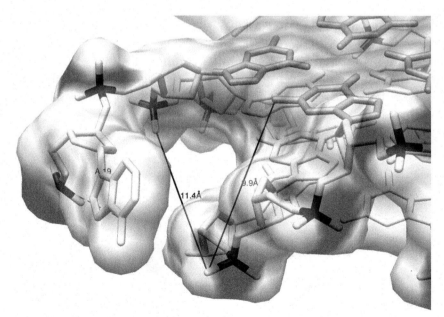

FIGURE 3–15 A view into a groove of the 22-mer crystal structure, now seen from above the groove and showing the solvent-accessible surface with 50% transparency. The adenine (here A19) from a TTA loop is seen acting as a flap over part of the groove. The view also highlights the role played by this loop in forming part of one boundary of the groove, and shows the hollow region in the interior of the groove. Groove dimensions are defined by distance between phosphate groups, and between one phosphate group and guanine substituent N2 at the floor of the groove.

The two G-quartet ends of the 22-mer quadruplex are not identical. The 5′ G-quartet (Figure 3–16a) is less polar and marginally more exposed, with a larger surface area than the 3′ G-quartet (Figure 3–16b) as a consequence of the protrusion of the four deoxyribose groups at the end of each strand onto the surface. The planar, accessible nature and the size of these surfaces suggest that they are very suitable for interaction with planar aromatic molecules, a theme that will be explored in detail in Chapter 6. Similarly the groove/loop cavities are in principle at least accessible for small molecule binding.

NMR Structures of Human Telomeric Quadruplexes in Potassium Solution

Early NMR studies on human telomeric sequences in potassium ion solution did not result in interpretable spectra corresponding to a single species of an intramolecular quadruplex, but instead suggested the presence of multiple equilibria (summarized by Phan, 2009). This is in accord with some interpretations of the biophysical data on four-repeat human telomeric sequences in K^+ solution, at least at the mM concentrations usual in NMR experiments (see Section 3.5 below). Analysis of the bimolecular two-repeat sequence d(TAGGGTTAGGGT) similarly showed the presence of several forms (Phan & Patel, 2003). However, it was also found that the equilibrium could be shifted to a particular topology by judicious choice of substitution of a thymine by a uracil or a 5-bromouracil nucleoside. (This is closely analogous to the strategy adopted in order to successfully crystallize a given sequence—"ringing the changes" with it being empirically found that variations in thymine→5-bromouracil mutations in particular can enhance crystal diffracting quality and even the ability of a sequence to crystallize at all.) The NMR analysis of the 12-mer in dilute (3 mM) solution with the sequence having a uracil

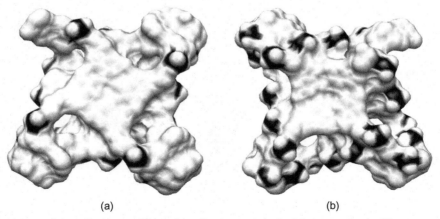

(a)   (b)

FIGURE 3–16 Views of the solvent-accessible surfaces for the two terminal G-quartets in the 22-mer crystal structure, showing **(a)** the 5′ end and **(b)** the 3′ end. The surfaces of the phosphorus atoms in the backbones are shown colored black.

substitution at position six resulted in a single species with parallel topology, closely similar to the bimolecular crystal structure. This was the first unequivocal evidence that the all-parallel form of a human telomeric quadruplex can exist in dilute solution, outside of the crystalline state. On the other hand, 5-bromouracil substitution at the 7-position resulted in the equilibrium favoring an anti-parallel structure with two lateral loops, one at each end of the quadruplex.

A very distinct topology has been found in a number of subsequent NMR studies in dilute solution. This, the anti-parallel (3 + 1) quadruplex topology (Figure 3–17) is so called on account of the distinctive pattern of three backbones being oriented in one direction and the fourth in the other, which is a consequence of a pattern of three *syn/anti* and one *anti/syn* glycosidic angles at each G-quartet (Zhang, Phan & Patel, 2005). It has been observed with several human telomeric sequences in solution (Ambrus *et al.*, 2006; Luu *et al.*, 2006; Dai *et al.*, 2007b), and was also independently proposed on the basis of a CD and melting study of the 22-mer sequence with systematic substitutions of 8-bromoguanine for guanine (Xu, Noguchi & Sugiyama, 2006). The initial observation of the (3 + 1) form was in an asymmetric structure containing a three-repeat sequence associating with a separate single-repeat sequence (Zhang, Phan & Patel, 2005), although subsequent studies have focused on four-repeat intramolecular quadruplex sequences. For example, the sequence d[$T_2G_3(T_2AG_3)_3A$] forms almost entirely a single species in K^+ solution, with the NMR structure revealing the characteristic (3 + 1) intramolecular arrangement (Luu *et al.*, 2006) having a single propeller loop followed by two lateral loops, one at each end of the quadruplex (Figure 3–18a). The (3 + 1) topology was also observed (Dai *et al.*, 2007b) in the solution structure formed by the sequence d[$A_3G_3(T_2AG_3)_3AA$], which highlights the important stabilizing role of flanking sequences, forming here an adenine base-triplet with one adenine from the 5′-stem and one each from the chain-reversal loop and a lateral loop. These two sequences differ slightly at the 5′ and 3′ ends from fully natural human telomeric ones, but a subsequent NMR study (Phan *et al.*, 2007) has shown that the (3 + 1) motif can still be formed by completely natural sequences such as d[$TAG_3(T_2AG_3)_3$]. In all these NMR structures the flanking sequences are still performing active roles in adding extra stacking stability to the ends.

Lateral loop

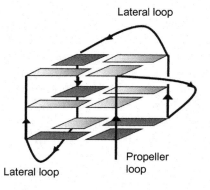

Lateral loop   Propeller loop

FIGURE 3–17 Schematic view of the (3 + 1) quadruplex topology, as found in NMR structures (Zhang, Phan & Patel, 2005; Ambrus *et al.*, 2006; Luu *et al.*, 2006). The individual rectangles indicate the glycosidic conformation at each guanosine nucleoside; light shading indicates an *anti* conformation and darker shading a *syn* one.

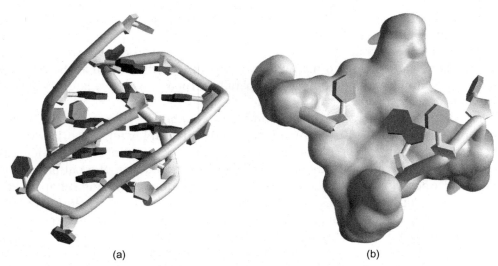

(a)   (b)

FIGURE 3–18 (a) The structure of one of the structures taken from an ensemble of (3 + 1) NMR structures, in PDB entry id AQ7 (Zhang, Phan & Patel, 2005), drawn in cartoon form. **(b)** shows one end of this structure, with the lateral loop and the TTA bases in cartoon form and the terminal G-quartet in solvent-accessible surface representation.

The sensitivity of telomeric sequences to undergo polymorphic change has been further shown by the discovery of the effect of small changes to the 3' end of some flanking sequences that can form (3 + 1) structures. An example is the addition of TT, as in d[TAG$_3$(T$_2$AG$_3$)$_3$TT], which results in the dominance of a second and distinctive (3 + 1) form, the so-called hybrid-2 topology (Figure 3–19a, b) in which the polarity is reversed so that the chain-reversal loop now follows the two lateral loops (Phan, Luu & Patel, 2006). This terminal sequence requirement has also been found in the quadruplex structure of d[(T$_2$AG$_3$)$_4$TT], with the TT end participating in a stabilizing TAT triplet at one end of the quadruplex (Dai *et al.*, 2007b), as well as being a feature in the structure of d[TAG$_3$(T$_2$AG$_3$)$_3$TT] (Phan *et al.*, 2007).

Another and more surprising form, with just two G-quartet layers, has more recently been reported, on the basis of an NMR analysis in K$^+$ solution of the natural 22-mer human telomeric sequence d[(GGGTTA)$_3$GGGT]. This structure (Lim *et al.*, 2009b) has an overall basket appearance with an anti-parallel topology, two lateral loops and one diagonal loop (Figure 3–20a, b). It thus resembles the fold found for the anti-parallel 22-mer in Na$^+$ solution (Wang & Patel, 1993), with one less G-quartet layer and with two loops extended in length by the addition of some of the extra guanosines. Although the NMR experiments did not locate the K$^+$ ion in the two-layer structure, it can be assumed that there would only be a single ion present, equidistant between the two layers. A two-quartet structure would in principle be less stable than a three-layer one, but in this case there is extensive compensating base stacking and base–base hydrogen-bonding interactions involving the lateral loop bases and the four guanines not involved in the G-quartets, with an A·G·A base triplet formed between two lateral loops and stacked

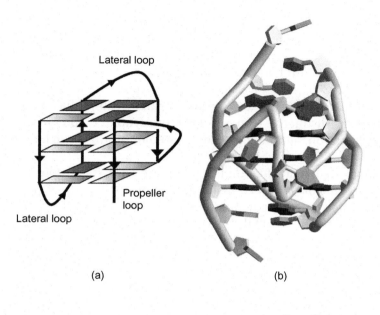

(a)

(b)

FIGURE 3–19 (a) Schematic view of the (3 + 1) hybrid-2 quadruplex topology (Dai *et al.*, 2007a). Shading of base representations is as in Figure 3.17. **(b)** Cartoon representation of a (3 + 1) hybrid-2 structure (PDB id 2JPZ) (Dai *et al.*, 2007b).

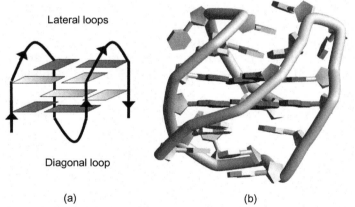

(a)

(b)

FIGURE 3–20 (a) Schematic view of the two-quartet basket quadruplex topology (Lim *et al.*, 2009b). Shading of base representations is as in Figure 3.17. **(b)** Cartoon representation of the basket structure (PDB id 2KF8) (Lim *et al.*, 2009a). The two central G-quartets are shaded.

onto a G-quartet. On the other side a G·G·G triplet is formed between the diagonal loop and the guanine close to the 3' end, and stacks onto the other G-quartet. The sequence d[A(G$_3$TTA)$_2$G$_3$TTAG$_3$T] (Zhang *et al.*, 2010) has multiple conformations in solution; however, replacing one guanosine with an inosine (i.e. d[A(G$_3$TTA)$_2$IG$_2$TTAG$_3$T]) resulted in a single species with the same two G-quartet basket topology. These authors also noted that this fold could not be assigned on the basis of CD spectra alone, and suggest that it may represent an intermediate between other folds.

Sequence variation in human telomeric DNA does occur (Allshire, Dempster & Hastie, 1989), though it is rare. The structural consequences of such changes have been explored in an NMR study of a sequence (Table 3–1) (Lim *et al.*, 2009a), where three TTAGGG repeats have been replaced by CTAGGG. This sequence still forms an anti-parallel intramolecular quadruplex (Figure 3–21a), but with three lateral loops and an unusual asymmetric G·C·C·G quartet (Figure 3–21b) stacked on two normal G-quartets at one end of the core of the structure. This quartet has three strong hydrogen bonds, within each G·C base pair, with only a loose association between them; a somewhat more symmetric and firmly hydrogen bonded G·C·C·G quartet was found in the solution structure of a fragile X syndrome triplet repeat quadruplex (Kettani, Kumar & Patel, 1995). It has been suggested (Lim *et al.*, 2009a) that the distinctive features of this quadruplex could form a platform for ligand-selective recognition, although to date no studies on this have been reported.

The observation of six distinct topologies for human telomeric DNA quadruplexes does suggest that further exploration of other sequence possibilities will uncover yet more folds. However, their relevance to longer telomeric sequences and in particular to the human telomeric overhang when in a cellular environment is unclear. The presence of the stabilizing base–base and base stacking interactions involving flanking sequences and both lateral and strand-reversal loops in all of the (3 + 1) structures and the two-layer basket structure contributes to their stability, and may explain why the parallel form is not observed with (3 + 1)-forming sequences. Analogous stabilization (and thus the (3 + 1) motif) is not obviously available to the original 22-mer sequence $d[AG_3(T_2AG_3)_3]$ since the necessary flanking sequences are absent. The precise nature of any non-parallel species formed by $d[AG_3(T_2AG_3)_3]$ in relatively dilute K^+ solution therefore remains to be determined, with evidence that a number of forms coexist under these conditions. A very recent determination by NMR methods (Heddi & Phan, 2011) of the solution structures for the original 22-mer and several other telomeric sequences

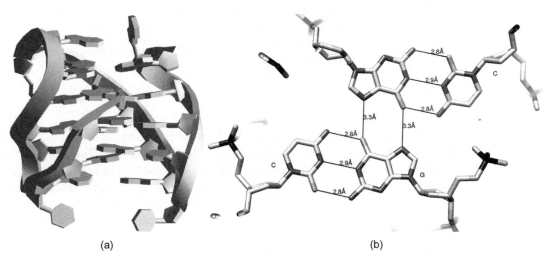

(a)   (b)

FIGURE 3–21 **(a)** Cartoon representation of the quadruplex formed by $d[A(G_3CTA)_3G_3]$ (Lim *et al.*, 2009a). **(b)** Detailed view of the G·C·C·G quartet in one of the NMR structures from the ensemble of deposited structures, showing the hydrogen-bond distances.

in K$^+$ solution but under conditions of molecular crowding (PEG 200) has provided new insights into solution topologies. This analysis found that a single species is formed for all of the sequences, and that they all have the parallel fold topology. Detailed analysis has shown that these structures are essentially identical, apart from some minor differences in preferred loop conformations, to the structure originally observed in the crystalline state (Parkinson, Lee & Neidle, 2002). This NMR analysis thus supports the concept that the parallel topology is present under physiological conditions, whereas more dilute solution favors a complex equilibrium of several forms.

Biophysical Studies on Telomeric Quadruplex Topology

The technique of circular dichroism (CD) has been frequently used to discriminate between different quadruplex folds, with over 100 publications (to May 2011) using CD to analyze telomeric quadruplexes. CD spectral features derive from particular patterns of base stacking interactions. The basis of many CD quadruplex structural assignments is the underlying change in G-quartet stacking between anti-parallel and parallel structures (see, for example, Hardin *et al.*, 1991; Petraccone *et al.*, 2005; Paramasivan, Rujan & Bolton, 2007). These reflect the effects of stacking differences on the different topologies caused by *anti* compared to *syn* glycosidic conformations. A positive CD band around 260 nm and a minimum at around 240 nm indicates a parallel quadruplex topology whereas a positive band around 290–295 nm with a negative band near 260 nm indicates an anti-parallel arrangement. More complex spectra can make it not possible to unambiguously assign structural type to a given quadruplex, especially if the loops also have stacked bases.

CD spectra in Na$^+$ solution for human intramolecular telomeric quadruplex sequences have used the Na$^+$ NMR structure (Wang & Patel, 1992) as an unambiguous basis for structural assignment, and concur that both techniques show basket-type anti-parallel topology. The situation in K$^+$ solution is altogether more complex (see, for example, Balagurumoorthy & Brahmachari, 1994; Rujan, Meleney & Bolton, 2005; Vorlíčková *et al.*, 2005; Ren iuk *et al.*, 2009), and there appears to be dependence on such factors as the number of telomeric repeats, flanking sequences and K$^+$ concentration. Under most conditions, sequences with four repeats appear to form multiple species in solution, with both parallel and anti-parallel forms being present. This interpretation is in accord with data from a range of biophysical methods including single-molecule fluorescence energy transfer (Ying *et al.*, 2003) and pulsed electron paramagnetic resonance using a spin-labeled telomeric sequence (Singh *et al.*, 2009). On the other hand, studies interpreting data from cross-linking or DNA backbone cleavage reagent experiments (Redon *et al.*, 2003; He, Neumann & Panyutin, 2004) suggest the dominance of a chair-type anti-parallel form (which has not been observed by NMR or crystallography). An anti-parallel structure close to that found by NMR in Na$^+$ solution has been proposed (Li *et al.*, 2005) on the basis of a comparison of experimental hydrodynamic data to sedimentation coefficients calculated from different structural models, with the conclusion that the crystallographically determined parallel structure for the human telomeric quadruplex is only relevant to the crystalline state. The subsequent NMR studies on a number of sequences, as discussed above, have revealed added complexity with the possibility of several distinct species being present (Antonacci, Chaires & Sheardy, 2007), their exact nature and ratio at equilibrium being dependent on flanking sequences.

Several recent studies have examined the effects of molecular crowding (Miyoshi, Nakao & Sugimoto, 2002) on human telomeric sequences in solution, and conclude that under these conditions, in contrast with the dilute solutions typically used in the studies outlined above, the parallel form becomes the dominant species present. Thus a combination of CD, gel electrophoresis and fluorescence data has shown (Xue *et al.*, 2007) that the sequence d[$G_3(T_2AG_3)_3$] in 150 mM K^+ and PEG 200 forms a parallel quadruplex, as in the crystalline state (Parkinson, Lee & Neidle, 2002). These conditions may be considered to approximate those in a cell, which has ~40% w/v of contents being macromolecules (Zimmerman & Trach, 1991); this is close to the ~50% of content in a quadruplex crystal being solvent. It was also found (Xue *et al.*, 2007) that crowding resulted in decreased processivity by telomerase, an effect ascribed to the increased stability of the parallel form. A transition to the parallel form can also be induced in some human telomeric sequences by the use of ethanol as an inducer of crowding (Renčiuk *et al.*, 2009), and significantly in the absence of any crowding reagent, by increasing the quadruplex concentration. At concentrations of 1 mM up to 40 mM, the CD spectra indicated an anti-parallel form, but above this level, and especially beyond 100 mM, no anti-parallel CD spectra were observable and instead the parallel form was the sole species present.

The Organization and Structure of Quadruplex Repeats along the Telomeric DNA Overhang

A number of experimental studies have addressed the question of the preferred number of quadruplex repeats along a long biologically relevant telomeric DNA repeat sequence, as well as their organization, most recently using near-molecular imaging methods. Atomic force microscopy has been used to examine a 96-mer (Xu *et al.*, 2009), and the images show four adjacent blob shapes arranged along an individual DNA molecule, consistent with them being adjacent G-quadruplexes. It is suggested that this can be represented by a model (Figure 3–22a) involving consecutive (3 + 1) units with stacking of the intervening TTA loops, similar to the conceptual model based directly on the (3 + 1) NMR structure (Ambrus *et al.*, 2006). A beads-on-a-string model (Yu, Miyoshi & Sugimoto, 2006) has taken into account the observations that long folded telomeric sequences have CD spectra indicative of both parallel and anti-parallel topologies. The thermodynamic data are suggestive of a model in which individual quadruplex units (which can be either parallel or anti-parallel), are connected by TTA loops, and are not stacked on each other. Since diagonal and lateral loops are flexible, the implication is that the resulting multimer will similarly be non-rigid. The AFM images are also consistent

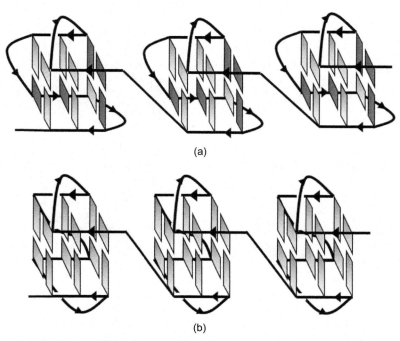

FIGURE 3–22 Two hypothetical models for the assembly of the telomeric single-stranded overhang into multimer quadruplexes. **(a)** Using the (3 + 1) hybrid quadruplex topology as the repeating unit, and **(b)** using the parallel topology.

with an alternative model in which adjacent parallel quadruplexes are held together by direct G-quartet–G-quartet stacking (Figure 3–22b)—at the level of resolution of the AFM images, it is not possible to discriminate between such models. Another AFM study (Wang *et al.*, 2011), also using a 96-mer human telomeric DNA sequence, by contrast, found that less than the maximum of four quadruplexes were normally formed, and that two is statistically the most likely number. Surprisingly this study also suggests that some quadruplexes can still remain on the 96-mer when molecules of the single-strand binding protein POT1 are also bound.

A molecular modeling study using 10 ns molecular dynamics simulations has examined models of two consecutive quadruplexes, and has compared a range of possible models with experimental data from CD and hydrodynamic data, using established algorithms (Li *et al.*, 2005) to compare predicted sedimentation coefficients from the various models with experimental observations, on the basis of the differences in molecular dimensions between models (Petraccone, Trent & Chaires, 2008). This analysis found that the stacked (3 + 1) hybrid and parallel quadruplex dimers were the most stable in the simulations, but that only the dimensions of the former were in accord with the experimental hydrodynamic data. The individual monomer quadruplexes in the dimer model examined in this study are connected by loose base stacking through adjacent lateral

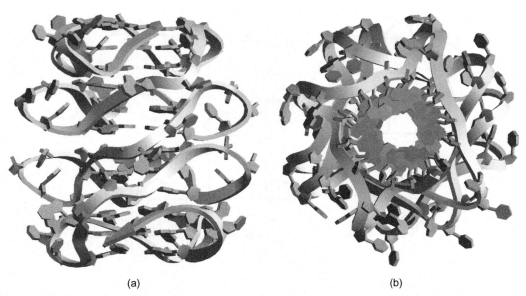

(a)   (b)

FIGURE 3–23 The modeled structure of a four quadruplex-repeat multimer based on the parallel telomeric quadruplex crystal structure, following a molecular dynamics simulation study, **(a)** showing the quasi-helical nature of the arrangement, and the TTA loops connecting each unit. **(b)** A view down the helix axis, i.e. the central cation channel. Coordinates kindly supplied by Dr. S.M. Haider, Queen's University, Belfast.

loops. By contrast a detailed molecular dynamics analysis of multimers, built up from individual parallel quadruplexes as the repeating units (Haider, Parkinson & Neidle, 2008), has arranged the monomer units to be stacked on one another through direct G-quartet contact, and linked by external TTA linker loops in a distinctive manner, acting as hinges between quadruplexes (Figure 3–23a). This linkage gives some flexibility between the multimer units, allowing movement of individual units as suggested by the data of Yu, Miyoshi & Sugimoto (2006). The TTA linker loops can also potentially act as recognition motifs in view of their external accessibility.

The contrasting features of these models await resolution by future crystallographic or NMR studies, but as discussed earlier in this chapter, in the context of single quadruplex units, it may well be the case that the diverse models are not irreconcilable, and simply reflect distinct solution/cellular conditions. Thus, although conformational complexity is commonly observed for single quadruplex units in dilute solution such as that used in conventional NMR or CD experiments, more concentrated solution (and the crowded environment of the cell nucleus) would favor more densely packed arrangements for the telomeric DNA overhang such as those of stacked parallel quadruplexes. The facile manner in which the parallel architecture can potentially self-aggregate was first noted when the parallel crystal structure was determined (Parkinson, Lee & Neidle, 2002). This aggregation results in a helical-like arrangement (Figure 3–23b) that has similarities to the original 5′-GMP and poly (G) helices discussed in Chapter 2.

Conclusions

Studies aimed at determining the preferred overall folds of human telomeric quadruplexes have developed into a minor industry over the past nine years. It is now becoming apparent that the perceived differences in derived structural types are real, but that they reflect distinct experimental conditions: (i) sequence changes, notably flanking sequences, are capable of altering the preferred topology from one anti-parallel form to another, in large part by the stacking influence of the flanking sequences on relative stabilities, and (ii) concentration of telomeric DNA can more profoundly affect topological preferences, as can crowding agents, whether they are really promoting crowding or just acting as dehydrating reagents (Miller *et al.*, 2010). Thus dilute solution and crystal environments are not the same, and the latter may well be the more physiologically relevant of the two, although still an idealization of the greater complexity of the biological molecule, the complete telomeric single-stranded overhang.

References

Alberti, P., Mergny, J.-L., 2003. DNA duplex-quadruplex exchange as the basis for a nanomolecular machine. Proc. Natl. Acad. Sci. USA 100, 1569–1573.

Allshire, R.C., Dempster, M., Hastie, N.D., 1989. Human telomeres contain at least 3 types of G-rich repeat distributed non-randomly. Nucleic Acids Res. 17, 4611–4627.

Ambrus, A., Chen, D., Dai, J., Bialis, T., Jones, R.A., Yang, D., 2006. Human telomeric sequence forms a hybrid-type intramolecular G-quadruplex structure with mixed parallel/antiparallel strands in potassium solution. Nucleic Acids Res. 34, 2723–2735.

Antonacci, C., Chaires, J.B., Sheardy, R.D., 2007. Biophysical characterization of the human telomeric (TTAGGG)$_4$ repeat in a potassium solution. Biochemistry 46, 4654–4660.

Balagurumoorthy, P., Brahmachari, S.K., 1994. Structure and stability of human telomeric sequence. J. Biol. Chem. 269, 21858–21869.

Collie, G.W., Parkinson, G.N., Neidle, S., Rosu, F., De Pauw, E., Gabelica, V., 2010. Electrospray mass spectrometry of telomeric RNA (TERRA) reveals the formation of stable multimeric G-quadruplex structures. J. Amer. Chem. Soc. 132, 9328–9334.

Dai, J., Carver, M., Punchihewa, C., Jones, R.A., Yang, D., 2007. Structure of the hybrid-2 type intramolecular human telomeric G-quadruplex in K$^+$ solution: insights into structure polymorphism of the human telomeric sequence. Nucleic Acids Res. 35, 4927–4940.

Dai, J., Punchihewa, C., Ambrus, A., Chen, D., Jones, R.A., Yang, D., 2007. Structure of the intramolecular human telomeric G-quadruplex in potassium solution: a novel adenine triple formation. Nucleic Acids Res. 35, 2445–2450.

Duquette, M.L., Handa, P., Vincent, J.A., Taylor, A.F., Maizels, N., 2004. Intracellular transcription of G-rich DNAs induces formation of G-loops, novel structures containing G4 DNA. Genes Dev. 18, 1618–1629.

Haider, S., Parkinson, G.N., Neidle, S., 2008. Molecular dynamics and principal components analysis of human telomeric quadruplex multimers. Biophys. J. 95, 296–311.

Hardin, C.C., Henderson, E., Watson, T., Prosser, J.K., 1991. Monovalent cation induced structural transitions in telomeric DNAs: G-DNA folding intermediates. Biochemistry 30, 4460–4472.

He, Y., Neumann, R.D., Panyutin, I.G., 2004. Intramolecular quadruplex conformation of human telomeric DNA assessed with ^{125}I-radioprobing. Nucleic Acids Res. 32, 5359–5367.

Heddi, B., Phan, A.T., 2011. Structure of human telomeric DNA in crowded solution. J. Amer. Chem. Soc. DOI: 10.1021/ja200786q

Kettani, A., Kumar, R.A., Patel, D.J., 1995. Solution structure of a DNA quadruplex containing the fragile X syndrome triplet repeat. J. Mol. Biol. 254, 638–656.

Kuryavyi, V., Majumdar, A., Shallop, A., Chernichenko, N., Skripkin, E., Jones, R., et al., 2001. A double chain reversal loop and two diagonal loops define the architecture of a unimolecular DNA quadruplex containing a pair of stacked G(syn)-G(syn)-G(anti)-G(anti) tetrads flanked by a G-(T-T) Triad and a T-T-T triple. J. Mol. Biol. 310, 181–194.

Lai, M., Podell, E.R., Cech, T.R., 2004. Structure of human POT1 bound to telomeric single-stranded DNA provides a model for chromosome end-protection. Nature Struct. Mol. Biol. 11, 1223–1229.

Lane, A.N., Chaires, J.B., Gray, R.D., Trent, J.O., 2008. Stability and kinetics of G-quadruplex structures. Nucleic Acids Res. 36, 5482–5515.

Li, J., Correia, J.J., Wang, L., Trent, J.O., Chaires, J.B., 2005. Not so crystal clear: the structure of the human telomere G-quadruplex in solution differs from that present in a crystal. Nucleic Acids Res. 33, 4649–4659.

Lim, K.W., Alberti, P., Guédin, A., Lacroix, L., Riou, J.-F., Royle, N.J., et al., 2009a. Sequence variant (CTAGGG)$_n$ in the human telomere favors a G-quadruplex structure containing a G·C·G·C tetrad. Nucleic Acids Res. 37, 6239–6248.

Lim, K.W., Amrane, S., Bouaziz, S., Xu, W., Mu, Y., Patel, D.J., et al., 2009b. Structure of the human telomere in K$^+$ solution: a stable basket-type G-quadruplex with only two G-tetrad layers. J. Amer. Chem. Soc. 131, 4301–4309.

Luu, K.N., Phan, A.T., Kuryavyi, V., Lacroix, L., Patel, D.J., 2006. Structure of the human telomere in K$^+$ solution: an intramolecular (3 + 1) G-quadruplex scaffold. J. Amer. Chem. Soc. 128, 9963–9970.

Miller, M.C., Buscaglia, R., Chaires, J.B., Lane, A.N., Trent, J.O., 2010. Hydration is a major determinant of the G-quadruplex stability and conformation of the human telomere 3′ sequence of d(AG$_3$)(TTAG$_3$)($_3$)). J. Amer. Chem. Soc. 132, 17105–17107.

Miyoshi, D., Nakao, A., Sugimoto, N., 2002. Molecular crowding regulates the structural switch of the DNA G-quadruplex. Biochemistry 41, 15017–15024.

Nandamukar, J., Cech, T.R., Podell, E.R., 2010. How telomeric protein POT1 avoids RNA to achieve specificity for single-stranded DNA. Proc. Natl. Acad. Sci. USA 107, 651–656.

Neaves, K.J., Huppert, J.L., Henderson, R.M., Edwardson, J.M., 2009. Direct visualization of G-quadruplexes in DNA using atomic force microscopy. Nucleic Acids Res. 37, 6269–6275.

Neidle, S., 2007. Principles of nucleic acid structure. Academic Press, San Diego.

Neidle, S., Parkinson, G.N., 2008. Quadruplex DNA crystal structures and drug design. Biochimie 90, 1184–1196.

Paramasivan, S., Rujan, I., Bolton, P.H., 2007. Circular dichroism of quadruplex DNAs: applications to structure, cation effects and ligand binding. Methods 43, 324–331.

Parkinson, G.N., Lee, M.P.H., Neidle, S., 2002. Crystal structure of parallel quadruplexes from human telomeric DNA. Nature 417, 876–880.

Patel, D.J., Phan, A.T., Kuryavyi, V., 2007. Human telomere, oncogenic promoter and 5′-UTR G-quadruplexes: diverse higher order DNA and RNA targets for cancer therapeutics. Nucleic Acids Res. 35, 7429–7455.

Petraccone, L., Erra, E., Esposito, V., Randazzo, A., Galeone, A., Barone, G., et al., 2005. Biophysical properties of quadruple helices of modified human telomeric DNA. Biopolymers 77, 75–85.

Petraccone, L., Trent, J.O., Chaires, J.B., 2008. The tail of the telomere. J. Amer. Chem. Soc. 120, 16530–16532.

Phan, A.T., Mergny, J.-L., 2002. Human telomeric DNA: G-quadruplex, i-motif and Watson-Crick double helix. Nucleic Acids Res. 30, 4618–4625.

Phan, A.T., Patel, D.J., 2003. Two-repeat human telomeric d(TAGGGTTAGGGT) sequence forms inter-converting parallel and antiparallel G-quadruplexes in solution: distinct topologies, thermodynamic properties, and folding/unfolding kinetics. J. Amer. Chem. Soc. 125, 15021–15027.

Phan, A.T., 2009. Human telomeric G-quadruplex: structures of DNA and RNA sequences. FEBS J. 277, 1107–1117.

Phan, A.T., Kuryavyi, V., Luu, K.N., Patel, D.J., 2007. Structure of two intramolecular G-quadruplexes formed by natural human telomere sequences in K^+ solution. Nucleic Acids Res. 35, 6517–6525.

Phan, A.T., Luu, K.N., Patel, D.J., 2006. Different loop arrangements of intramolecular human telomeric (3 + 1) G-quadruplexes in K^+ solution. Nucleic Acids Res. 34, 5715–5719.

Redon, S., Bombard, S., Elizondo-Riojas, M.A., Chottard, J.-C., 2003. Platinum cross-linking of adenines and guanines on the quadruplex structures of the $AG_3(T_2AG_3)_3$ and $(T_2AG_3)_4$ human telomere sequences in Na^+ and K^+ solutions. Nucleic Acids Res. 31, 1605–1613.

Ren iuk, D., Kejnovská, I., Školáková, P., Bednářová, K., Motlová, J., Vorlíčková, M., 2009. Arrangements of human telomere DNA quadruplex in physiologically relevant K^+ solutions. Nucleic Acids Res. 37, 6625–6634.

Risitano, A., Fox, K.R., 2003. Stability of intramolecular DNA quadruplexes: comparison with DNA duplexes. Biochemistry 42, 6507–6513.

Rujan, I.N., Meleney, C., Bolton, P.H., 2005. Vertebrate telomere repeat DNAs favor external loop propeller quadruplex structures in the presence of high concentrations of potassium. Nucleic Acids Res. 33, 2022–2031.

Singh, V., Azarkh, M., Exner, T.E., Hartig, J.S., Drescher, M., 2009. Human telomeric quadruplex conformations studied by pulsed EPR. Angew. Chem. Int. Ed. Engl. 48, 9728–9730.

Tran, P.L.T., Mergny, J.-L., Alberti, P., 2011. Stability of telomeric G-quadruplexes. Nucleic Acids Res. 39, 3282–3294.

Vorlícková, M., Chládková, J., Kejnovská, I., Fialová, M., Kypr, J., 2005. Guanine tetraplex topology of human telomere DNA is governed by the number of (TTAGGG) repeats. Nucleic Acids Res. 33, 5851–5860.

Wang, H., Nora, G.J., Ghodke, H., Opresko, P., 2011. Single molecule studies of physiologically relevant telomeric tails reveals pot1 mechanism for promoting G-quadruplex unfolding. J. Biol. Chem. 286, 7479–7489.

Wang, Y., Patel, D.J., 1993. Solution structure of the human telomeric repeat d[AG3(T2AG3)3] G-tetraplex. Structure 1, 263–282.

Xu, Y., Ishizuka, T., Kurabayashi, K., Komiyama, M., 2009. Consecutive formation of G-quadruplexes in human telomeric-overhang DNA: a protective capping structure for telomere ends. Angew. Chem. Int. Ed. Engl. 48, 7833–7836.

Xu, Y., Noguchi, Y., Sugiyama, H., 2006. The new models of the human telomere $d[AG_3(T_2AG_3)_3]$ in K^+ solution. Bioorg. Med. Chem. 14, 5584–5591.

Xue, Y., Kan, Z.Y., Wang, O., Yao, Y., Liu, J., Hao, Y.H., et al., 2007. Human telomeric DNA forms parallel-stranded intramolecular G-quadruplex in K+ solution under molecular crowding conditions. J. Amer. Chem. Soc. 129, 11185–11191.

Ying, L., Green, J.J., Li, H., Klenerman, D., Balasubramanian, S., 2003. Studies on the structure and dynamics of the human telomeric G-quadruplex by single-molecule fluorescence resonance energy transfer. Proc. Natl. Acad. Sci. USA 100, 14629–14634.

Yu, H.-Q., Miyoshi, D., Sugimoto, N., 2006. Characterization of structure and stability of long telomeric DNA G-quadruplexes. J. Amer. Chem. Soc. 128, 15461–15468.

Zhang, N., Phan, A.T., Patel, D.J., 2005. (3 + 1) assembly of telomeric repeats into an asymmetric dimeric G-quadruplex. J. Amer. Chem. Soc. 127, 17277–17285.

Zhang, Z., Dai, J., Veliath, E., Jones, R.A., Yang, D., 2010. Structure of a two-G-tetrad intramolecular G-quadruplex formed by a variant human telomeric sequence in K$^+$ solution: insights into the inter-conversion of human telomeric G-quadruplex structures. Nucleic Acids Res. 38, 1009–1021.

Zheng, K.W., Chen, Z., Hao, Y.H., Tan, Z., 2010. Molecular crowding creates an essential environment for the formation of stable G-quadruplexes in long double-stranded DNA. Nucleic Acids Res. 38, 327–338.

Zhou, J., Wei, C., Jia, G., Wang, X., Feng, Z., Li, C., 2009. Human telomeric G-quadruplex formed from duplex under near physiological conditions: spectroscopic evidence and kinetics. Biochimie 91, 1104–1111.

Zimmerman, S.B., Trach, S.O., 1991. Estimation of macromolecule concentrations and excluded volume effects for the cytoplasm of *Escherichia coli*. J. Mol. Biol. 222, 599–620.

4

Telomeric Quadruplex Ligands I: Anthraquinones and Acridines

Introduction

The finding that low molecular weight quadruplex-binding compounds could inhibit telomerase catalytic function has been followed over the succeeding 14 years by a large number of studies on a wide range of small molecules, whose aim has been in the majority of instances to discover potential anti-cancer agents. Few of these compounds have to date progressed beyond biochemical evaluations, but many have provided important data on structure–activity relationships and on quadruplex-small molecule recognition. There are many reviews available that cover some of the material discussed in this chapter: those by Monchaud & Teulade-Fichou (2008), Tan, Gu & Wu (2008), Yang & Okamoto (2010) are especially notable. The field has not only grown but diversified, and to best reflect these many studies, the present and following chapters focus on telomeric quadruplex ligands, with succeeding chapters in turn discussing ligand binding to RNA, genomic quadruplexes, and the biology of telomeric quadruplex ligands.

Ethidium bromide (Figure 4–1a) was historically the first compound to be shown to interact with G-rich DNA sequences (Guo *et al.*, 1992) although the nature of the possible higher-order structures, the basis of the binding, or telomerase inhibition, were not investigated. A pioneering *in silico* search for quadruplex ligands (Chen, Kuntz & Shafer, 1996) was used to find the carbocyanine derivative DODC (Figure 4–1b) from a library of compounds (Figure 4–2), which was also shown to bind to a simple dimeric hairpin G-quadruplex in solution, although the exact binding mode (groove or intercalation was suggested), was not determined. Competition dialysis experiments (Ren & Chaires, 1999) have shown that both DODC and ethidium actually bind to duplex and triplex DNA with greater affinity, an early example of the need to consider selectivity between different nucleic acid structure types.

The issue of defining the ligand binding site(s) in a quadruplex has been answered by biophysical, crystallographic and NMR studies, which have been unanimous in demonstrating that, regardless of the exact nature of the ligand, the primary binding site is external to the central core of stacked G-quartets, on one or other of the terminal G-quartet faces (Figure 4–1c). There is no evidence of intercalative binding analogous to that in duplex DNA (Figure 4–1d). This is understandable in view of the high stability of quadruplex structures, which suggests that the flexibility of a duplex containing stacked

Therapeutic Applications of Quadruplex Nucleic Acids. DOI: 10.1016/B978-0-12-375138-6.00004-2

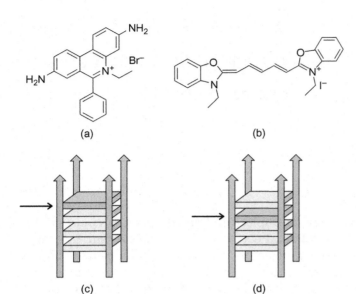

FIGURE 4–1 (a) Structure of ethidium bromide. **(b)** Structure of the carbocyanine derivative DODC (Chen, Kuntz & Shafer, 1996). **(c)** Schematic of a highlighted ligand, shown as a shaded rectangular solid, bound onto the end G-quartet (unshaded rectangular solid) of a G-quadruplex. **(d)** An alternative binding mode, with the ligand intercalated between two adjacent G-quartets.

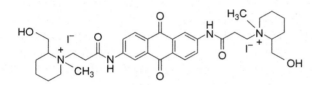

FIGURE 4–2 Structure of the disubstituted anthraquinone compound 2,6-bis[3-[2-(2-hydroxymethyl)piperidino]propionamido] anthracene-9,10-dione.

Watson-Crick base pairs is much more than in a quadruplex with a core of three or four stacked G-quartets (Olsen, Gmeiner & Marky, 2006).

Methods for Studying Quadruplex–Ligand Complexes

The determination of telomerase inhibition by telomeric quadruplex ligands continues to be a central feature of many studies. Almost all use one or other variants of the TRAP assay protocol, sometimes employing commercial kits. Very few laboratories are set up to run direct telomerase assays, so results from different laboratories may not be comparable and, as discussed in Chapter 1, issues of ligand crossover into the PCR step of the TRAP assay may compromise some data. However, results from the literature are presented here "as is" since trends may be apparent even if quantitative comparisons are not always meaningful. Measurements of cell growth inhibition are also subject to variation due to individual laboratory practice and assay type chosen.

The evaluation of quadruplex-small molecule binding has used a number of biophysical methods, many of which had been of proven value in earlier studies of ligand binding to duplex DNA (see, for example, Lane *et al.*, 2008; Fox, 2010). All of these methods are sensitive to ionic strength and buffer conditions, especially when applied to human telomeric quadruplexes with their structural dependence on a K^+ concentration of at least 50 mM. This again means that detailed comparisons between results from different laboratories (and sometimes from the same laboratory!) need to be made with care.

Among the most commonly used methods applied to quadruplexes are:

1. CD methods, with the emphasis on examining changes in topology. These can enable overall quadruplex topologies to be assigned, but are not always reliable since this requires unambiguous correlation with a known structure at the outset (see Lane *et al.*, 2008, for a detailed discussion of these and other issues).

2. UV/visible/fluorescence spectroscopy to analyze equilibrium binding and obtain binding constants (Mergny, Phan & Lacroix, 1998), using, for example, Scatchard plot analysis of ligand binding to a quadruplex. This method is rarely used since it is slow, uses large quantities of oligonucleotide and it is not straightforward to adequately deal with the complexities of multiple binding sites.

3. Melting methods to examine effects analogous to the classic temperature-dependent helix $\leftrightarrow$ coil transition of duplex nucleic acids, typically monitored by changes in absorbance at 260 nm, which increase on going from duplex to single strand. A sharp melting transition (the T_m) is also typically observed when the temperature of a solution containing a quadruplex sequence is raised, whose value in degrees is indicative of the relative stability of a particular quadruplex (Rachwal & Fox, 2007). A significant hyperchromic shift at 295 nm is unique to G-quadruplex structures (Mergny, Phan & Lacroix, 1998), and thus its observation is taken as definitive evidence of the formation of a quadruplex structure. A typical T_m value for a human telomeric four-repeat sequence would be ~60°. Addition of a ligand will increase the T_m if it stabilizes the structure, with this being signified as the ΔT_m value, which varies with relative concentration of ligand and DNA. It is common to quote the ΔT_m value at a standard ligand concentration, such as 1 μM, when comparing effects from different ligands. CD changes on melting can also be used to obtain ΔT_m values.

4. A commonly employed variant on the melting technique is based on the FRET (Fluorescence Resonance Energy Transfer) principle. A quadruplex oligonucleotide is labeled at the 5′ and 3′ ends with, respectively, a donor fluorophore (for example, FAM, 6-carboxyfluorescein) and an acceptor (for example, TAMRA, 6-carboxytetramethylrhodamine) (Mergny & Maurizot, 2001; Guédin, Lacroix & Mergny, 2010). When the two are in close proximity in solution as when the sequence is folded into a quadruplex the fluorescence emission from the donor is quenched whereas there is a large increase in the donor emission signal as the temperature is raised. In common with UV melting behavior a sharp transition is typically observed, and the midpoint of the transition is taken as the T_m value. The technique is highly sensitive, with a large change in signal so that only low concentrations of oligonucleotide and ligand are needed. The technique is highly amenable to high-throughput methods, and commonly uses a real-time PCR instrument that only

needs minimal adaptation for this purpose, together with a 96-well plate format (Darby *et al.*, 2002; Guyen *et al.*, 2004; Schultes *et al.*, 2004). One potential problem with the FRET approach is that ligand binding to donor or acceptor fluorophore can occur, although the number of recorded instances of this are low (Guédin, Lacroix & Mergny, 2010). The FRET method is also well suited to the study of quadruplex–ligand selectivity. A common approach is to perform ΔT_m determinations in the presence of increasing molar ratios of a second nucleic acid—calf thymus DNA is often used to assess selectivity for a quadruplex versus duplex DNA. The approach can equally be used to examine selectivity for a particular quadruplex over others by using excess molar ratios of other quadruplexes.

5. Equilibrium binding and kinetic data can be obtained using the real-time technique of surface plasmon resonance (SPR). There is increasing availability of automated SPR instruments, some of which can be used in a medium throughput manner. SPR provides data of particular use for understanding ligand recognition, on equilibrium-binding constants (k_a), stoichiometry (n, number of ligand molecules bound) and Gibbs free energy (ΔG). The SPR technique is based on the detection of differences in flow characteristics that can be sensed by optical techniques when a solution of ligand is flowed over a surface on which the target macromolecules (in this instance quadruplex oligonucleotides) are immobilized (Liu & Wilson, 2010). A particular advantage of SPR is its ability to work with low concentrations of quadruplex oligonucleotide and to provide accurate assessments of strong binding.

6. Isothermal titration calorimetry (ITC) is increasingly being used to provide accurate thermodynamic data (Lane *et al.*, 2008). It also provides an important check on the validity of binding parameters from other methods such as SPR, although the inherently low throughput of ITC means that it tends to be used on a few selected compounds in a series and not as a screening method.

7. Electrospray ionization mass spectrometry (ESI-MS) is a gas-phase method for analyzing molecular species and their association/dissociation. Its successful application to quadruplexes and their ligand complexes is relatively recent (Rosu, De Pauw & Gabelica, 2008; Gabelica, 2010), but several studies have shown that it is capable of providing data on association constants that compares well with that from other techniques (for example, Lombardo *et al.*, 2010). It does require the absence of metal ions such as K^+ or Na^+, so it normally performed in the presence of NH_4^+ ions, which do stabilize quadruplexes (although remarkably few studies have been reported on the detailed effects on quadruplex structure of changing from K^+ or Na^+).

8. Footprinting methods employ chemical cleavage agents whose patterns of DNA strand break are altered in the presence of a ligand bound at a specific position in a quadruplex structure (Sun & Hurley, 2010). Chemical protection can thus help to determine preferred sites of binding, and is most powerful when information is available about the folding of the targeted quadruplex. A common reagent used is dimethyl sulfate, which binds to N7 of guanine, and thus disrupts G-quartet structures. The porphyrin compound TMPyP4 can be used as a self-cleavage reagent since it produces light-induced cleavage at sites close to where a molecule is bound.

Anthraquinones, their Derivatives and Analogs

Substituted amido-anthraquinones were the first class of molecules to be shown to have activity as quadruplex-mediated telomerase inhibitors (Zahler *et al.*, 1991; Sun *et al.*, 1997), and the first non-nucleoside inhibitors of telomerase catalytic function to be studied in depth. The basic anthraquinone core is common to several natural and synthetic clinically useful anti-cancer agents (Adriamycin, mitoxantrone) that bind to duplex DNA and earlier studies had shown that chemical manipulation of substituted anthraquinones could result in compounds with selectivity for triple-stranded DNA (Fox *et al.*, 1995). Initial proof of principle studies were undertaken on the disubstituted quaternary compound (Figure 4–2), which has a $^{tel}IC_{50}$ value of 56 μM for the extent of telomerase enzymatic inhibition, as measured using a direct assay. This method provides a reliable estimate of inhibition since it is not subject to the problems and uncertainties of the less direct TRAP assay (see Chapter 1). The binding of this compound to quadruplex nucleic acids was demonstrated spectrophotometrically, although the interactions were not quantitated. The choice of amido linkage was based on the assumption that extending the planar aromatic surface of the chromophore would enhance π–π stacking interactions with a planar G-quartet—no molecular structures of ligand–quadruplex complexes were available but data on the few native quadruplex structures were used for qualitative molecular modeling. Subsequent studies (Perry *et al.*, 1998a, b; Neidle *et al.*, 2000) systematically examined the effects of varying the substitution pattern around the anthraquinone ring system on telomerase inhibition and on cancer cell growth, as well as on a limited subset of compounds for quadruplex binding using ITC (Read *et al.*, 1999) (Figure 4–3). A representative set of compounds in the 1,4, 1,5, 1,8, 2,6 and 2,7 series were evaluated and the effects of differing side-chains were examined. This established that as well as the anthraquinone core itself, side-chains terminating in cationic end-groups (Figure 4–4) are important contributors to quadruplex binding and to the extent of telomerase inhibition. A set of structure–activity relationships was derived from the telomerase inhibitory data:

1. Two side-chains are optimal for activity.
2. The optimal side-chain length is $-(CH_2)_{2\text{-}3}-$.
3. The amido group is important.
4. Piperidine or pyrrolidine cationic end-groups are optimal.
5. There is a progressive drop in activity as the size of the end-group is increased from a six-membered ring to an eight-membered ring and then bulky bicyclic substituents.

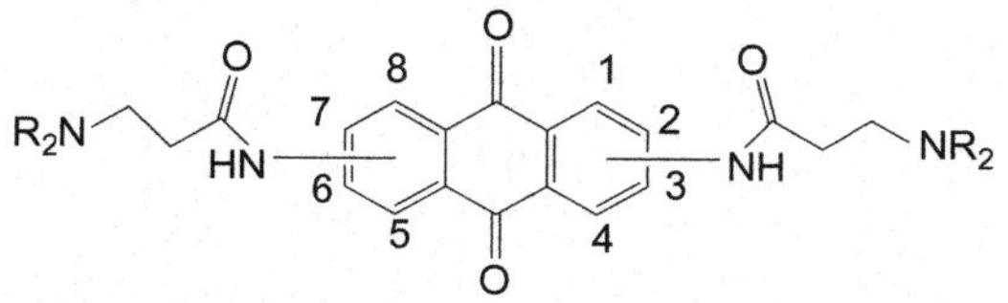

FIGURE 4–3 General structure of amidoalkylamino disubstituted anthraquinone compounds.

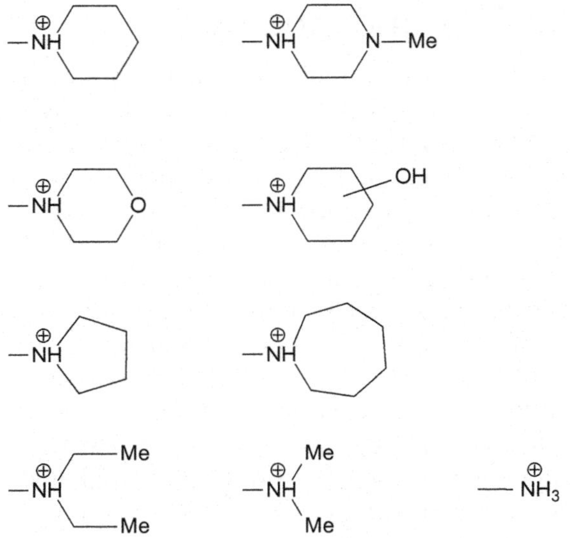

FIGURE 4–4 The preferred cationic amine end-groups used for many quadruplex-binding ligands.

6. The cationic charge is important. Compounds with, for example, morpholino groups are generally inactive.

7. The pattern of ring substitution is of only secondary importance. This is undoubtedly a consequence of the ability of the regioisomers of these disubstituted compounds to adopt a broadly similar range of geometries when bound to a quadruplex (Figure 4–5).

These early anthraquinone studies attempted to rationalize the available data with molecular modeling based on the most relevant experimentally determined molecular structure then available, that from the NMR study of a human telomeric 22-mer in Na^+ solution (Wang & Patel, 1993). It was hypothesized that binding would not occur by classic intercalation, but instead the concept of end-stacking was proposed, in which the ligand molecule stacks on to the terminal G-quartet in a quadruplex structure, which would be more energetically favorable than the unstacking of adjacent G-quartets required in an intercalation model (Read *et al.*, 1999; Read & Neidle, 2002). Subsequent crystallographic, NMR and biophysical data have confirmed the correctness of this assumption. QSAR approaches (Zambre *et al.*, 2010) have supported the requirement of cationic side-chains, using the available biological data to generate a predictive model.

No experimental structure of a synthetic anthraquinone–quadruplex complex has been reported, but a crystal structure of the anti-cancer antibiotic daunomycin with the tetramolecular quadruplex formed from four strands of d(TGGGGT) has been determined (Clark *et al.*, 2003). This drug has an anthraquinone core together with a charged glucosamine substituent attached to a fourth cyclohexene ring (Figure 4–6). The structure shows three daunomycin molecules stacked onto a terminal G-quartet of an individual quadruplex (Figure 4–7a), with this assembly forming a head-to-head dimer in the

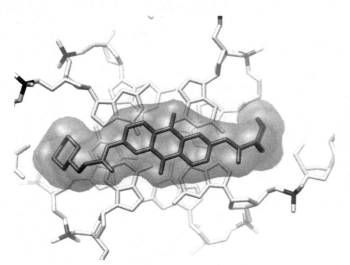

FIGURE 4–5 Molecular model of a 2,6-disubstituted amidoalkylamino anthraquinone bound in a low-energy position onto a G-quartet surface. The solvent-accessible surface of the ligand is also shown. Similar positions are found for a range of disubstituted anthraquinones.

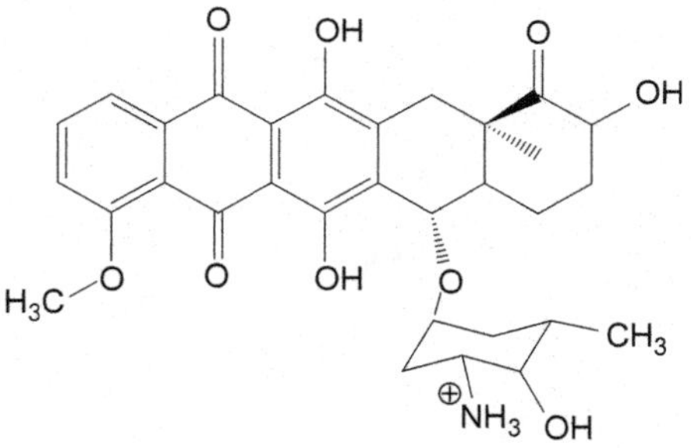

FIGURE 4–6 Structure of Adriamycin (doxorubicin). Daunomycin has the –CH₂OH group replaced by –CH₃.

unit cell such that there is a sandwich of three drug molecules stacking onto three more at the interface between the two quadruplexes. Remarkably there are no direct interactions between each of the three daunomycin molecules in each plane, and their overlap with individual guanines in the G-quartet is moderate. Instead, the substituent glucosamine rings fit snugly into the grooves (Figure 4–7b), and in two grooves there is direct hydrogen bonding between the substituent protonated amine group and a phosphate oxygen atom (Figure 4–7c). This is a rare instance of a cationic side-chain–phosphate interaction being observed in a quadruplex–ligand crystal structure and suggests that amino-substituted sugar groups may be useful in future ligand design. A recent study (Manet *et al.*, 2011) of the closely related drug Adriamycin (doxorubicin) binding to a human telomeric quadruplex sequence suggests rather different stoichiometry, 1:1 and 2:1, which may reflect the differences between more complex intramolecular quadruplexes with TTA loops compared to the simpler tetramolecular quadruplex used in the crystal structure analysis.

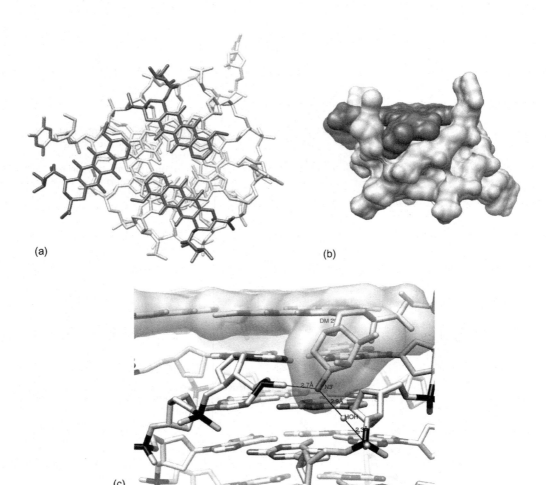

FIGURE 4–7 Views from the crystal structure of the d(TGGGGT)–daunomycin complex (PDB id 1OOK: Clark *et al.*, 2003). **(a)** View projected onto the terminal G-quartet, with the three bound daunomycin molecules shown with shaded stick bonds. **(b)** A side view into one of the grooves, with the structure drawn in solvent-accessible surface mode and daunomycin molecules dark shaded. A substituent daunosamine group is shown fitting into a quadruplex groove. **(c)** Detailed view of hydrogen-bond interactions in the groove involving the daunosamine amine group, a water molecule and phosphate groups.

A number of related 2,7-disubstituted amidoanthraquinones have been reported with significant activity in the NCI 60 cell line panel together with activity in the TRAP assay (Huang *et al.*, 2008). Anthraquinones with a range of peptide-like substituents at the ends of side-chains have been devised and both 2,6- and 2,7-disubstituted with terminal lysine and arginine groups show moderate quadruplex affinity and telomerase inhibition (Zagotto *et al.*, 2008a, b), with ΔT_m values of 5–8° and telomerase $^{tel}IC_{50}$ values in the low μM range. The majority of amidoanthraquinone compounds evaluated to data that display telomerase inhibitory activity also have generalized cellular toxicity,

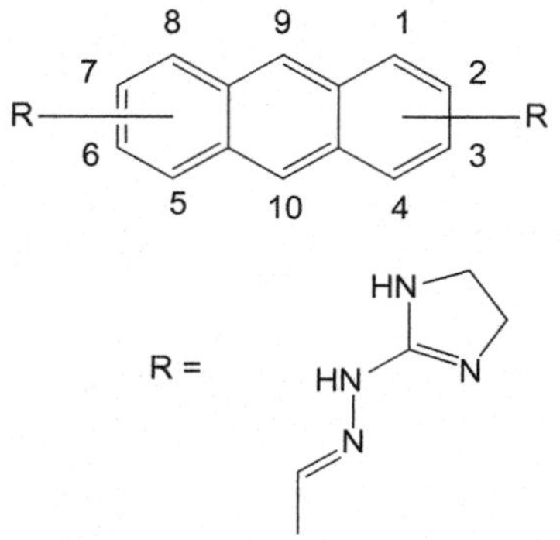

FIGURE 4–8 The anthracene core, with imidazole-containing side-chains substituents.

with typical IC_{50} values against laboratory cancer cell lines that are closely comparable, also in the low μM range. A series of anthracene derivatives (Figure 4–8) with imidazole-containing side-chains substituents (the parent compound bisantrene is an established duplex DNA binding molecule) has more selective behavior (Folini *et al.*, 2010). The 1,5-derivative in particular selectively stabilizes a human telomeric quadruplex sequence with a ΔT_m value of ~19° compared to a value of ~6° with a duplex DNA sequence and a 13-fold difference in K_a value. This compound is active in the telomerase TRAP assay with a $^{tel}IC_{50}$ value of 0.9 μM, comparing favorably with the IC_{50} value for its antiproliferative activity against several cancer cell lines averaging ~16 μM. Thus, by contrast with the classic amidoanthraquinones discussed above, this compound has a significant window in which to observe effects due to telomerase inhibition and telomere targeting followed by cell death.

Disubstituted Acridines

The tricyclic acridine chromophore was originally developed (Harrison *et al.*, 1999) as an alternative platform to the anthraquinones, in order to circumvent the poor aqueous solubility of many amidoanthraquinone derivatives. The central ring nitrogen atom in acridines such as 3,6-diaminoacridine (proflavine) is protonated at physiological pH (Albert, 1966), which would aid solubility as well as increase electron deficiency through the chromophore, leading to enhanced G–quadruplex interactions. This prediction was borne out in a series of 3,6-disubstituted amidoacridines (Figure 4–9), with side-chains closely similar to those established in the earlier amidoanthraquinone series. Their ability to inhibit the telomerase enzyme is approximately correlated with their antiproliferative activity and their relative quadruplex affinity relates, as with the anthraquinone derivatives, to the cationic charge and size of the end-groups on the side chains (Read *et al.*, 1999).

A crystallographic analysis of a complex between the bimolecular *Oxytricha nova* 12-mer quadruplex (sequence d[GGGGTTTTGGGG]) and the bis-pyrrolidino compound 7 has been reported. The structure (Haider, Parkinson & Neidle, 2003) reveals a ligand molecule bound at one end of the quadruplex, stacked onto the terminal G-quartet of the core, and inserted in between this quartet and the T4 diagonal loop (Figure 4–10).

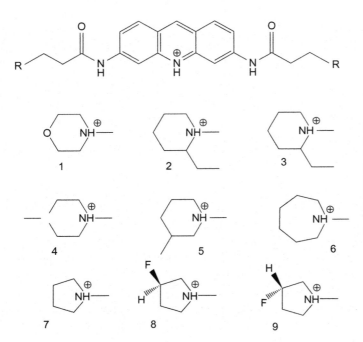

FIGURE 4–9 The general structure of 3,6-diamidoalkylaminoacridines, with the central ring nitrogen atom protonated in solution at physiological pH. The various amine end-groups that have been explored crystallographically (Haider, Parkinson & Neidle, 2003; Campbell *et al.*, 2009, 2011) are also shown.

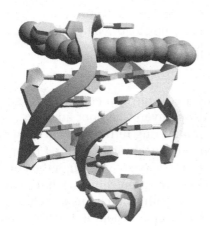

FIGURE 4–10 View of the crystal structure (Haider, Parkinson & Neidle, 2003) between bimolecular *Oxytricha nova* 12-mer quadruplex d[GGGGTTTTGGGG] and the bis-pyrrolidino compound 7, drawn in cartoon mode with the ligand in space-filling representation.

The loop conformation has had to be reorganized in order to accommodate the ligand, and one of the thymines has swung in to be in-plane with the acridine ring, with which it forms a pair of tight hydrogen bonds (Figure 4–11). The hydrogen bond between the thymine O2 atoms and the acridine central ring nitrogen atom is of especial significance since it provides definitive proof that this nitrogen atom is fully protonated at the physiological-like buffer conditions used in the crystallization.

A series of subsequent crystal structures with varying end-groups on the side-chains (Table 4–1) has explored the effects of size on binding-site geometry (Campbell *et al.*, 2009, 2011). The diagonal loop in this quadruplex enables a large binding cavity to be formed which can accommodate a wide range of end-groups. For example, a methyl group attached to a piperidino ring (compounds 3–5) can adopt a variety of orientations that reflect its position on the ring. It is striking that these varied orientations are not accompanied by any changes in loop geometry, which contrasts with what occurs with parallel loops (see below), where the steric requirements are much more stringent and loop conformational changes are the norm. It thus appears that a wide range of ligand geometries and side-chains can be accommodated in a diagonal loop binding site. At the same time the particular features of an amidoacridine favor a thymine-containing diagonal loop, in view of the stabilizing effect of the in-plane hydrogen bonding (Figure 4–11). It is conceivable that a loop cytosine, albeit in a distinct orientation, could hydrogen bond to an amidoacridine in an analogous manner.

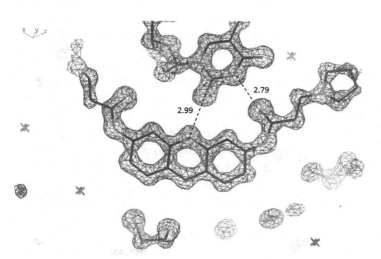

FIGURE 4–11 A detailed view of the structure of an *Oxytricha nova* quadruplex-acridine complex, projected onto the plane of the acridine group (PDB 3NZ7).

Table 4–1 Some details of the crystal structures of disubstituted acridines complexed with the *Oxytricha nova* quadruplex. the individual ligand end-groups are shown in Figure 4.9. Two distinct crystal forms were found for compound 2, resulting in two crystal structures, 2a and 2b

Ligand	1	2a	2b	3	4	5	6	7	8
PDB id	3EM2	3EQW	3EUI	3ERU	3ES0	3ET8	3EUM	3NPY	3NZ7
Resoln (Å)	2.30	2.20	2.20	2.00	2.20	2.45	1.78	1.18	1.10
No. water mols	64	66	159	71	56	51	52	176	187

Other types of disubstituted acridines have been reported. Several members of a series of 4,5-bis(dialkylaminoalkyl) acridines, i.e. with no amido substituents, show impressively potent telomerase inhibition (Laronze-Cochard et al., 2009), with one compound (Figure 4–12) having a $^{tel}IC_{50}$ value of 0.15 μM. Generally in this series quadruplex stabilization as measured by a FRET assay was only moderate with the best compound having a ΔT_m value of 3.0° at a 1 μM ligand concentration. Both quadruplex affinity and TRAP activity increased with increasing chain length. It is likely that the modest ΔT_m values reflect the lack of planarity of the sp^3 carbon atoms immediately attached to the acridine ring, which would decrease acridine-G-quartet overlap and thus diminish the effectiveness of any π–π interactions. It is unlikely that the acridine ring itself is protonated in these 4,5-compounds since the 4,5-dimethyl acridine precursor has a pK_a of 4.56 compared to 9.65 for 3,6-diaminoacridine (Albert, 1966), although no experimental pK_a determinations have been reported for them. So a significant diminution of quadruplex affinity compared to the 3,6 compounds is to be expected, although the unexpectedly high potency of telomerase inhibition shown by some of the 4,5-compounds suggests the involvement of non-quadruplex mediated mechanisms of action. Several members of a series of 4,5-disubstituted acridones (Figure 4–13), by contrast, although having very high quadruplex affinity and selectivity, do not have any activity in either a direct telomerase or a TRAP assay (Cuenca et al., 2009). Their ability to produce long-term growth arrest in cancer cell lines may be related to action at telomeres since senescent effects were observed, although other related mechanisms of action cannot be discounted, for example involving DNA damage induction at telomeres (see Chapter 6).

Another series of substituted acridines lacking a protonated ring nitrogen atom has extended triazole substituents at the 3- and 6-positions (Sparapani et al., 2010), generated by click chemistry. The substitutions with triazole rings were conceived to enhance coplanarity and stacking onto a G-quartet. Derivatives with pyrrolidino or diethylamino end-groups attached to $-(CH_2)_2-$ side-chains (Figure 4–14) were the most effective at telomeric quadruplex stabilization, and show selectivity for this particular quadruplex over either duplex DNA or several other quadruplex types. This selectivity is in qualitative agreement with the predictions from computer modeling. The overall structural arrangement has been

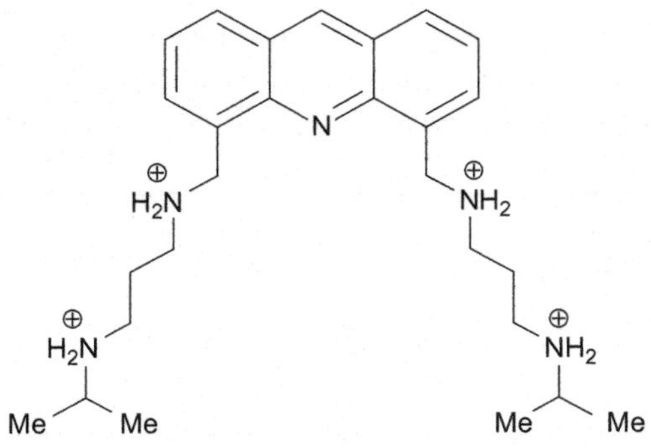

FIGURE 4–12 Structure of 4,5-bis(dialkylaminoalkyl) acridine.

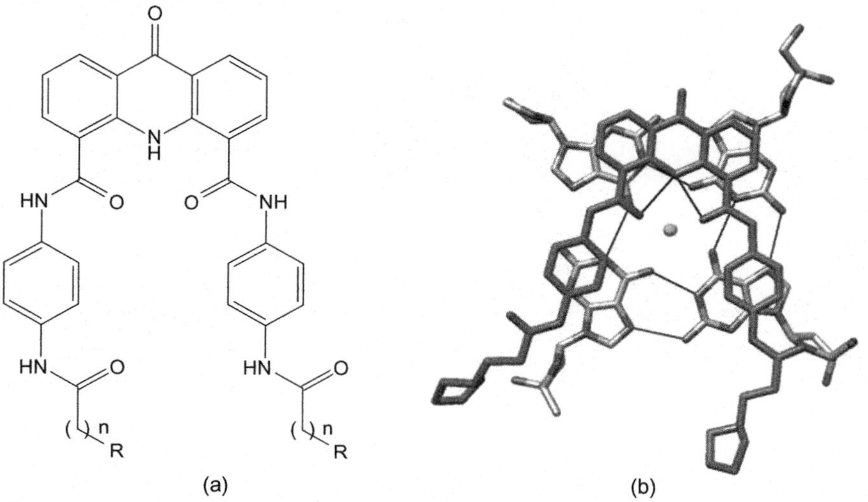

FIGURE 4–13 (a) General structure of 4,5-bisamidoarylalkylamino compounds. **(b)** Molecular model illustrating a possible overlay between a 4,5-bisamidoarylalkylamino compound and a G-quartet.

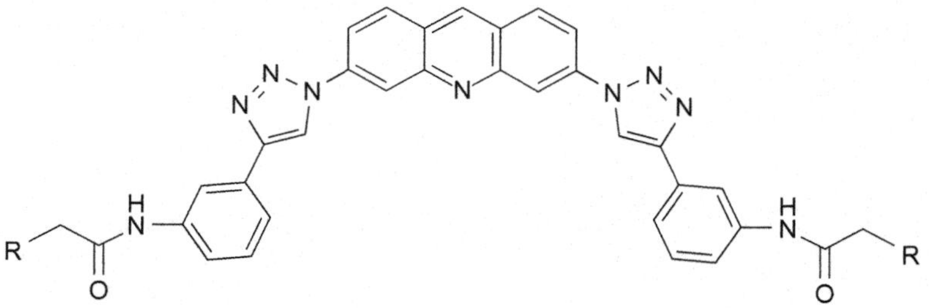

FIGURE 4–14 The click chemistry bis-triazole acridine [N,N'-((1,1'-(acridine-3,6-diyl)bis(1H-1,2,3-triazole-4,1-diyl)))] family of compounds. By contrast with the 3,6-amido acridines, the central ring nitrogen atom is not protonated.

revealed in a low-resolution (2.90 Å) crystal structure (Collie *et al.*, 2011) of a complex with a bimolecular human telomeric quadruplex, which confirms the original hypothesis (and rationale for synthesizing the compounds in the first instance), that the triazole rings are stacking over G-quartets (Figure 4–15). However, it is a reminder of the limitations of molecular modeling that the predicted position of the ligand (Figure 4–16) is about 1.5 Å from its crystallographically observed one. It is also notable that there are distinct orientations for the two side-chains in the experimental structure, which are necessary in order to avoid clashes with the loop. The best compound in the series has a relatively high ΔT_m value of 15.8° at a 1 μM ligand concentration, even though the central ring acridine nitrogen atom is not protonated, suggest that this factor is less important than the preservation and maximization of G-quartet stacking. Any direct substitution on the ring that counters this, such as of two sp³ carbon atoms, appears to have a large negative effect on quadruplex binding.

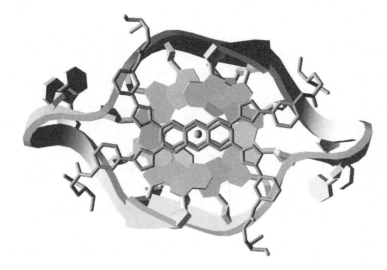

FIGURE 4–15 Cartoon of the 2.90 Å crystal structure (Collie *et al.*, 2011) of a bis-triazole acridine ligand complexed with a bimolecular human telomeric DNA quadruplex, viewed onto the acridine plane. The ligand is statistically disordered and both half-occupancy quadruplexes are shown here (PDB id 3QCR: Collie *et al.*, 2011). The quadruplex adopts a parallel topology.

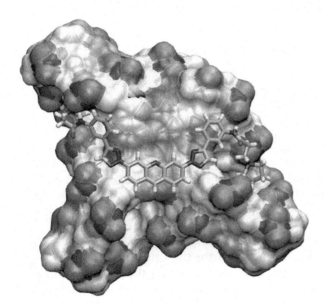

FIGURE 4–16 View of a molecular model for a bis-triazole acridine (in stick representation) bound to a human telomeric DNA quadruplex (shown in surface representation) (Sparapani *et al.*, 2010).

Trisubstituted Acridines

Following on from the realization that although the 3,6-disubstituted amidoacridines represented an improvement in telomerase potency, their selectivity and hence cellular window of activity was low, a systematic attempt was made to devise quadruplex ligands with improved selectivity. Molecular modeling was used (Read *et al.*, 2001), again taking the human telomeric 22-mer in Na^+ solution as a starting point. Selectivity was designed in by the simple expedient of having three substituents, based on the hypothesis that three would be able to interact in three quadruplex grooves, whereas since duplex DNA has only two grooves this feature would predict diminished binding to this form of DNA. The initial molecule synthesized, termed BRACO-19 (Figure 4–17), was a 3,6,9-trisubstituted acridine derivative with an anilino group at the 9-position. This group was previously exploited in the duplex DNA-binding acridine anti-cancer drug m-AMSA (amsacrine) that had been developed some years previously (Denny *et al.*, 1982) and has favorable electronic, steric and pharmacological properties. BRACO-19 shows a significant improvement in predicted and experimental quadruplex affinity and potency of telomerase inhibitory activity compared to its disubstituted acridine precursor, and also has a >10-fold difference in cell growth arrest in telomerase-expressing cancer cell lines (Table 4–2). The disubstituted acridine binds non-selectively to duplex and quadruplex DNA, whereas BRACO-19 has significant selectivity for the human telomeric quadruplex (with equilibrium binding constants determined by SPR).

Subsequent development of several series of BRACO-19 analogs focused on (i) a range of substituents in the 9-position, and (ii) changes in 3- and 6- substituent side-chain length and end-groups (Harrison *et al.*, 2003; Schultes *et al.*, 2004; Moore *et al.*, 2006; White *et al.*, 2007). The biology of BRACO-19 and its analogs is discussed in Chapter 6. The structure–activity relationships for these trisubstituted acridines can be summarized as:

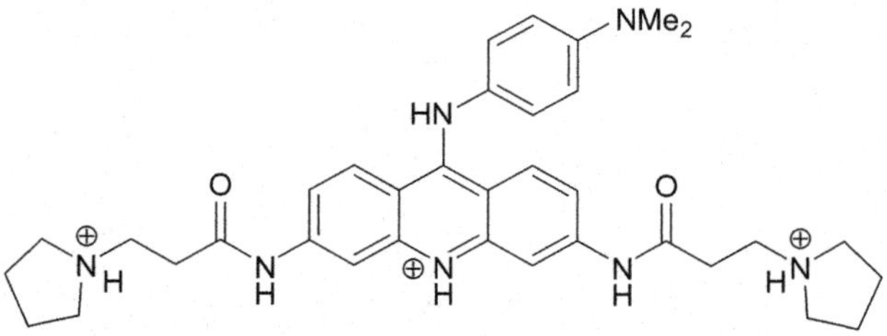

FIGURE 4–17 The structure of the trisubstituted acridine compound BRACO-19.

Table 4–2 Calculated (ΔE_{calc} in kcal mol^{-1}) and observed duplex/quadruplex affinities (K_a in mol^{-1}), together with telomerase inhibitory activity ($^{tel}IC_{50}$ in μM) and cell growth inhibition (IC_{50} in μM), for two cancer cell lines

Compound	ΔE_{calc}	Duplex K_a	Quad K_a	$^{tel}IC_{50}$	A2780	SKOV-3
Disubstituted acridine	−61.0	1.1×10^6	1.3×10^6	5.2	2.65	2.6
BRACO-19	−105.2	5.0×10^5	1.6×10^7	0.06	>25	>25

1. Uncharged 9-position substituents together with $-(CH_2)_2$- 3- and 6- side-chains results in both quadruplex binding and telomerase inhibition being optimal. Side-chains longer than $-(CH_2)_3$- lose quadruplex binding and telomerase potency.

2. Extending the 9-position substituent length does not have a major effect on either quadruplex binding or telomerase potency. The adverse effects of extended 3,6- substituents can be balanced by an extended substituent at the 9-position.

3. Protonation of the 9-position with extended side-chains results in elevated ΔT_m values and higher telomerase inhibitory potency.

4. Selectivity for quadruplex over duplex DNA is correlated with both affinity and telomerase inhibition.

An X-ray crystal-structure analysis at 2.5Å resolution of BRACO-19 complexed to a human bimolecular telomeric quadruplex (Campbell *et al.*, 2008) has provided a detailed view of how this ligand is bound. It is located at the interface between two separate quadruplexes, which together make up the "biological unit" in the crystal (Figure 4–18), with unusually the two quadruplexes oriented 3′ to 5′ to each other so that joining the two ends together would produce a continuous strand. These quadruplexes, in common with all other ligand-bound human telomeric quadruplex crystal structures determined to date, have a parallel fold, with propeller TTA loops. These have altered conformations compared to the loops in the native crystal structures (Parkinson, Lee & Neidle, 2002) and play a major role, in particular, in enclosing the ligand substituents into discrete binding pockets. One 3′ end thymine base is flipped into the binding site and becomes approximately coplanar with the acridine ring so that it interacts with the ring, in a

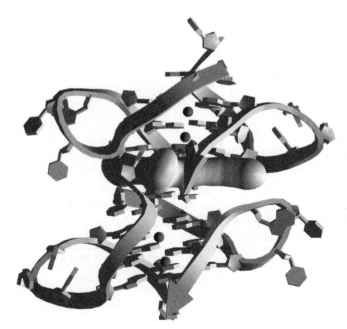

FIGURE 4–18 View of the biological unit of two bimolecular human telomeric quadruplexes and BRACO-19 complexed between them, as observed in the crystal structure (PDB id 3CE5: Campbell *et al.*, 2008). The solvent-accessible surface of the BRACO-19 molecule is shown.

manner analogous to that seen with the disubstituted acridines and the diagonal loops in the *Oxytricha* telomeric complexes (see Section 4.4). The difference here is that two water molecules play active bridging roles in thymine–acridine recognition (Figure 4–19). The amide functionality in both 3- and 6- side-chains is actively involved in this intricate hydrogen-bonding network, providing a rational basis for the structure–activity data that indicate their importance. The nature and dimensions of the loop binding pockets are also consistent with the structure–activity data (Harrison *et al.*, 2003; Schultes *et al.*, 2004; Moore *et al.*, 2006) on the length of the three side-chains and the size of their substituents. For example, the 9-anilino substituent fits into a narrow hydrophobic pocket at the quadruplex dimer interface (Figure 4–20). This pocket is open-ended so that although its dimensions favor phenyl and similar slim substituents at the acridine 9-position,

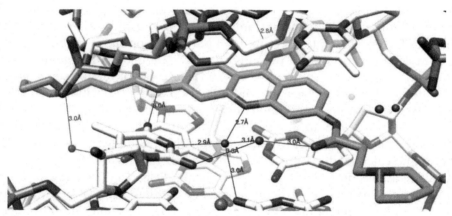

FIGURE 4–19 The BRACO-19 quadruplex complex: a detailed view of the interactions between the acridine ring nitrogen atom, two water molecules, a side-chain amido group and the in-plane thymine ring. A total of seven hydrogen bonds are involved.

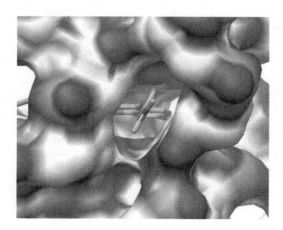

FIGURE 4–20 The BRACO-19 quadruplex complex: a view into one of the grooves where the 9-anilino group is bound. The ligand is shown in stick representation and the quadruplex as a solvent-accessible surface. Note that the orientation of the aniline group is well aligned with the shape of the surface.

extended para-substituents on the phenyl ring, such as long side-chains, would be allowed. Overall the BRACO-19 molecule utilizes seven out of eight possible hydrogen bonding opportunities, of which six are to water molecules.

It is also apparent from this and subsequent telomeric quadruplex–ligand crystal structures that the TTA loops play an active role in forming the ligand binding site, aided by the conformational flexibility of this trinucleotide loop. This means that the nature of the ligand itself determines the detailed structure of the binding site. So it is not sufficient to only consider G-quartet stacking when attempting to model these interactions, but the total geometry of the binding site needs to be modeled. The current small numbers of relevant experimental structures means that modeling of new ligand complexes is most likely to be successful when these are closely related to existing ones.

The extension of 9-position side-chain length in these trisubstituted acridines is at the expense of increased molecular weight even though the BRACO-19 crystal structure suggests that this would not have a deleterious effect on quadruplex affinity. Also there is some evidence that such molecules are less readily transported into cells. This was one important factor taken into account in an attempt to optimize the properties of the BRACO-19 molecule to produce a pre-clinical candidate telomere targeting compound (Martins *et al.*, 2007). This compound, AS1410 (Figure 4–21), retains the 3,6- substituents of the initial lead but replaces the anilino group at the 9-position with a fluorine-containing benzyl substituent, and modeling based on the crystal structure (Figure 4–22) suggests that this would be an acceptable modification. It was hypothesized that this would diminish the possibility of oxidative metabolism, although it would not improve lipophilicity (the calculated log P is 1.31 compared to 1.86 for BRACO-19—see Chapter 9 for a further discussion of log P applied to ligand design). AS1410 does stabilize the human telomeric quadruplex to a slightly greater extent than BRACO-19 with an increase in ΔT_m of 1.5° and has greater potency in the TRAP assay. The choice of this molecule for further development was based, at this point at least, on the assumption that high target affinity would be the most important factor defining *in vivo* potency. The biology of AS1410 is discussed in Chapter 6.

BRACO-19 has been the basis of several series of modified acridine compounds. Dimers linked at the 9-position by acyclic carbon chains are more selective for

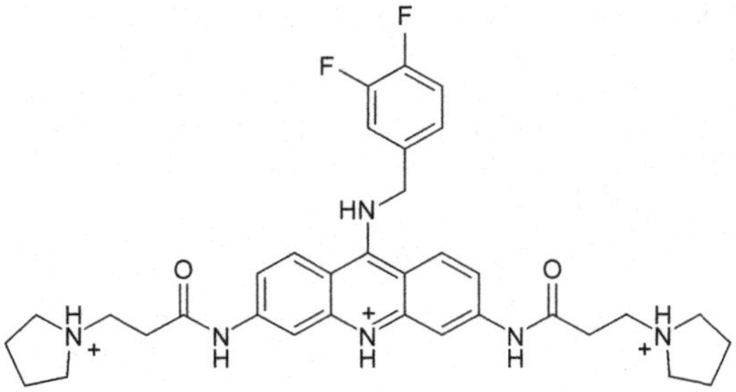

FIGURE 4–21 The structure of the trisubstituted compound AS1410, an analog of BRACO-19 (Martins *et al.*, 2007).

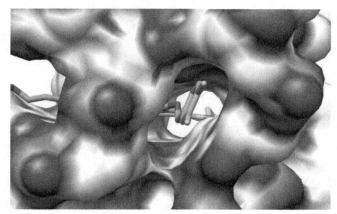

FIGURE 4–22 Computer modeling representation of AS1410 bound in the same position as observed in the BRACO-19 complex. The view is the same as in Figure 4.20. Note that the benzyl group does not appear to fit quite as well as the anilino group of BRACO-19, although the model has not been optimized.

quadruplex compared to duplex DNA (Fu *et al.*, 2009), although their quadruplex stabilization ability as estimated by T_m studies was only modestly superior to BRACO-19 itself. One compound is an exception and has –NH- groups in the linker, which would enhance electrostatic binding since these groups will be protonated at physiological pH. These dimers also show superior telomerase inhibition and inhibition of hPOT1 binding to a single-stranded overhang, although how they would interact with a quadruplex is not clear, at least on the basis of the BRACO-19 crystal structure.

The concept of extending the polycyclic acridine surface in order to enhance G-quartet stacking has been explored in a series of fused bispyrimidino-acridines (Figure 4–23), which appear to be high-affinity quadruplex binders, although the selectivity over duplex DNA was surprisingly enhanced in the absence of the 9-substituent (Debray *et al.*, 2009). These compounds could not bind to a quadruplex site in the same manner as BRACO-19 itself since the presence of the pyrimidino groups would preclude acridine–thymine contacts. Two separate studies have attempted to emulate the key three-substituent recognition feature of BRACO-19 more directly, and have also replaced the central acridine by a simple monocyclic aromatic group. In one an imidazole ring was selected as the core group from a 3D-QSAR analysis (Chen *et al.*, 2011) symmetrically substituted by three phenyl-N-methyl-piperazine groups (Figure 4–24a). A ΔT_m value of 23.5° for this molecule (using a FRET melting method) indicates high quadruplex affinity, even though the two proximal phenyl rings will necessarily be forced to deviate from coplanarity. A groove-binding mode for this molecule is feasible. Click chemistry has been used to assemble a tris-triazole compound (Figure 4–24b) with three cationic side-arms and a phenyl ring as core (Lombardo *et al.*, 2010). This compound can mimic the BRACO-19 binding mode (see Chapter 9) and also has higher quadruplex affinity and selectivity than its parent compound.

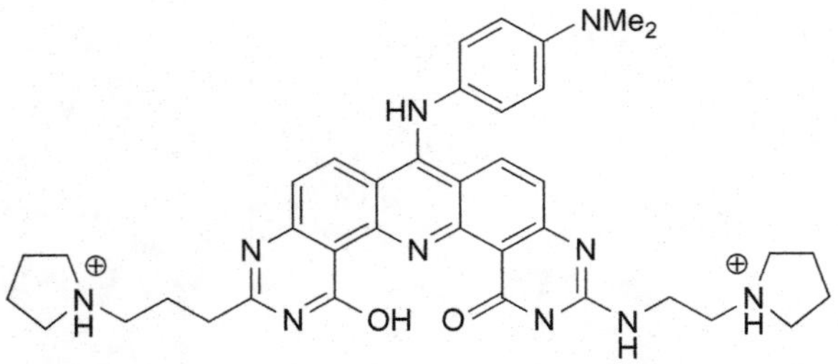

FIGURE 4–23 Structure of a polycyclic fused acridine mimetic of BRACO-19 (Debray *et al.*, 2009).

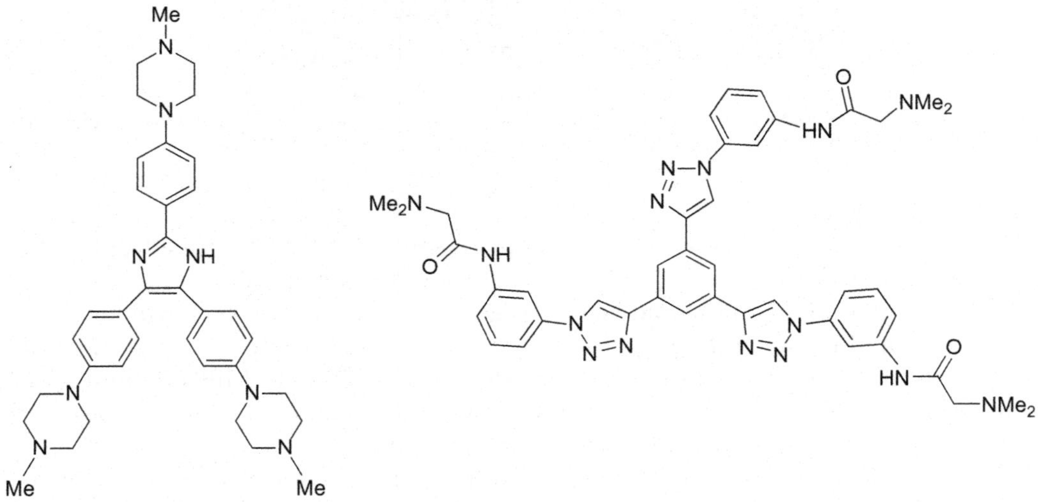

<table>
<tr><td>(a)</td><td>(b)</td></tr>
</table>

FIGURE 4–24 (a), (b) Structures of two non-polycyclic mimetics of BRACO-19, both with three extended side-chains.

Polycyclic Acridines—the RHPS4 Ligand

The concept of extending the size of the polycyclic portion of a quadruplex-binding ligand has clear attractions. It is a straightforward way of enhancing π–π stacking interactions with a terminal G-quartet. A series of quaternized quino[4,3,2-kl]acridinium salts has received considerable attention, with the lead compound RHPS4 (Figure 4–25) being established as a potent telomere targeting agent (Gowan *et al.*, 2001; Leonetti *et al.*, 2004). Its molecular pharmacology is discussed in Chapter 6. RHPS4 and a number of its analogs bind to human telomeric quadruplex DNA with high affinity (Cheng *et al.*, 2008) and some selectivity for quadruplex compared to duplex DNA, although the measured degree of selectivity is dependent on the methodology used. Surface plasmon resonance indicates 30-fold selectivity compared to five-fold from fluorescence titration measurements. The ΔT_m for RHPS4 at a 1 μM concentration using a FRET technique is 8.0°, apparently rather smaller than some other acridines, suggesting that the lack of cationic side-chains does result in a lower quadruplex stabilization ability, although this is not apparent in the biological profile of RHPS4. None of the 15 analogs examined in this study (Cheng *et al.*, 2008) have such side-chains, which may be reflected in their narrow range of telomerase inhibitory activity. Overall the initial molecule RHPS4 has the optimal mix of chemical and pharmacological features of any in the series, and has been the focus of a number of subsequent biological studies, which have shown significant antitumor activity, as well as elucidating details about its mode of action (Chapter 6).

RHPS4 prefers to bind to an anti-parallel human telomeric quadruplex in dilute solution, as judged by CD methods (Garner *et al.*, 2009). An NMR study of RHPS4 bound to the tetramolecular quadruplex d(TTAGGGT)$_4$, on the other hand, has shown that all four strands are parallel in this complex (Gavathiotis *et al.*, 2003). There are two ligand molecules bound per quadruplex (Figure 4–26), one at each end of the core of four stacked

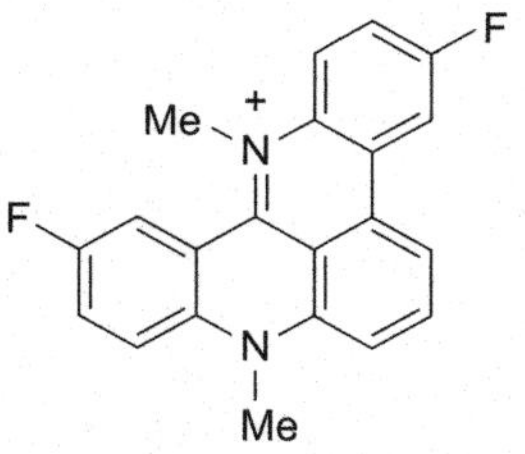

FIGURE 4–25 The structure of RHPS4.

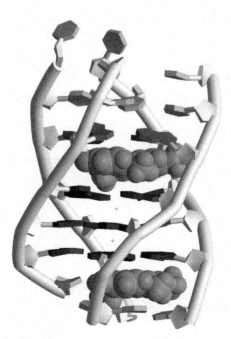

FIGURE 4–26 View of one of the ensemble of NMR models of the d(TTAGGGGT)$_4$ complex with RHPS4 (PDB id 1NMZ: Gavathiotis *et al.*, 2003), with ligand molecules shown in space-filling mode.

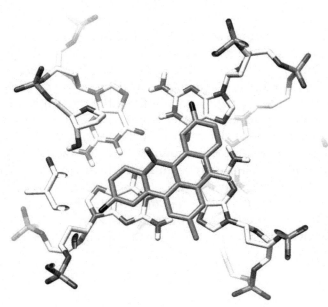

FIGURE 4–27 View onto the plane of the end-stacked RHPS4 molecule in the d(TTAGGGGT)$_4$ complex.

G-quartets. Regardless of the relevance of this particular structure to RHPS4 binding to human telomeric intramolecular quadruplexes, the G-quartet–ligand stacking (Figure 4–27) provides invaluable insights in principle into how the ligand can be modified to enhance affinity.

References

Albert, A., 1966. The Acridines, second ed. Edward Arnold, London.

Campbell, N.H., Parkinson, G.N., Reszka, A.P., Neidle, S., 2008. Structural basis of DNA quadruplex recognition by an acridine drug. J. Amer. Chem. Soc. 130, 6722–6724.

Campbell, N.H., Patel, M., Tofa, A.B., Ghosh, R., Parkinson, G.N., Neidle., S., 2009. Selectivity in ligand recognition of G-quadruplex loops. Biochemistry 48, 1675–1680.

Campbell, N.H., Smith, D.L., Reszka, A.P., Neidle, S., O'Hagan, D., 2011. Fluorine in medicinal chemistry: β-fluorination of peripheral pyrrolidines attached to acridine ligands affects their interactions with G-quadruplex DNA. Org. Biomol. Chem. 9, 1328–1331.

Chen, Q., Kuntz, I.D., Shafer, R.H., 1996. Spectroscopic recognition of guanine dimeric hairpin quadruplexes by a carbocyanine dye. Proc. Natl. Acad. Sci. USA 93, 2635–2639.

Chen, S.-B., Tan, J.-H., Ou, T.-M., Huang, S.-L., An, L.-K., Luo, H.-B., et al., 2011. Pharmacophore-based discovery of triaryl-substituted imidazole as new telomeric G-quadruplex ligand. Bioorg. Med. Chem. Lett. 21, 1004–1009.

Cheng, M.K., Modi, C., Cookson, J.C., Hutchinson, I., Heald, R.A., McCarroll, A.J., et al., 2008. Antitumor polycyclic acridines. 20. Search for DNA quadruplex binding selectivity in a series of 8,13-dimethylquino[4,3,2-kl]acridinium salts: telomere-targeted agents. J. Med. Chem. 51, 963–975.

Clark, G.R., Pytel, P.D., Squire, C.J., Neidle, S., 2003. Structure of the first parallel DNA quadruplex–drug complex. J. Amer. Chem. Soc. 125, 4066–4067.

Collie, G.W., Sparapani, S., Parkinson, G.N., Neidle, S., 2011. Structural basis of telomeric RNA quadruplex–acridine ligand recognition. J. Amer. Chem. Soc. 133, 2721–2728.

Cuenca, F., Moore, M.J., Johnson, K., Guyen, B., De Cian, A., Neidle, S., 2009. Design, synthesis and evaluation of 4,5-di-substituted acridone ligands with high G-quadruplex affinity and selectivity, together with low toxicity to normal cells. Bioorg. Med. Chem. Lett. 19, 5109–5113.

Darby, R.A., Sollogoub, M., McKeen, C., Brown, L., Risitano, A., Brown, N., et al., 2002. High throughput measurement of duplex, triplex and quadruplex melting curves using molecular beacons and a LightCycler. Nucleic Acids Res. 30, e39.

Debray, J., Zeghida, W., Jourdan, M., Monchaud, D., Dheu-Andries, M.-L., Dumy, P., et al., 2009. Synthesis and evaluation of fused bispyrimidinoacridines as novel pentacyclic analogues of quadruplex-binder BRACO-19. Org. Biomol. Chem. 7, 5219–5228.

Denny, W.A., Cain, B.F., Atwell, G.J., Hansch, C., Panthananickal, A., Leo, A., 1982. Potential antitumor agents. 36. Quantitative relationships between experimental antitumor activity, toxicity, and structure for the general class of 9-anilinoacridine antitumor agents. J. Med. Chem. 25, 276–315.

Folini, M., Pivetta, C., Zagotto, G., De Marco, C., Palumbo, M., Zaffaroni, N., et al., 2010. Remarkable interference with telomeric function by a G-quadruplex selective bisantrene regioisomer. Biochem. Pharmacol. 79, 1781–1790.

Fox, K.R. (Ed.), 2010. Drug–DNA Interaction Protocols. Methods in Molecular Biology. Humana Press, New Jersey.

Fox, K.R., Polucci, P., Jenkins, T.C., Neidle, S., 1995. A molecular anchor for stabilizing triple-helical DNA. Proc. Natl. Acad. Sci. USA 92, 7887–7891.

Fu, Y.T., Keppler, B.R., Soares, J., Jarstfer, M.B., 2009. BRACO19 analog dimers with improved inhibition of telomerase and hPot 1. Bioorg. Med. Chem. 17, 2030–2037.

Gabelica, V., 2010. Determination of equilibrium association constants of ligand–DNA complexes by electrospray mass spectrometry. Methods Mol. Biol. 613, 89–101.

Garner, T.P., Williams, H.E., Gluszyk, K.I., Roe, S., Oldham, N.J., Stevens, M.F., et al., 2009. Selectivity of small molecule ligands for parallel and anti-parallel DNA G-quadruplex structures. Org. Biomol. Chem. 7, 4194–4200.

Gavathiotis, E., Heald, R.A., Stevens, M.F., Searle, M.S., 2003. Drug recognition and stabilisation of the parallel-stranded DNA quadruplex d(TTAGGGT)$_4$ containing the human telomeric repeat. J. Mol. Biol. 334, 25–36.

Gowan, S.M., Heald, R., Stevens, M.F., Kelland, L.R., 2001. Potent inhibition of telomerase by small-molecule pentacyclic acridines capable of interacting with G-quadruplexes. Mol. Pharmacol. 60, 981–988.

Guédin, A., Lacroix, L., Mergny, J.-L., 2010. Thermal melting studies of ligand DNA interactions. Methods Mol. Biol. 613, 25–35.

Guo, Q., Lu, M., Marky, L.A., Kallenbach, N.R., 1992. Interaction of the dye ethidium bromide with DNA containing guanine repeats. Biochemistry 31, 2451–2455.

Guyen, B., Schultes, C.M., Hazel, P., Mann, J., Neidle, S., 2004. Synthesis and evaluation of analogues of 10H-indolo[3,2-b]quinoline as G-quadruplex stabilising ligands and potential inhibitors of the enzyme telomerase. Org. Biomol. Chem. 2, 981–988.

Haider, S.M., Parkinson, G.N., Neidle, S., 2003. Structure of a G-quadruplex–ligand complex. J. Mol. Biol. 326, 117–125.

Harrison, R.J., Cuesta, J., Chessari, G., Read, M.A., Basra, S.K., Reszka, A.P., et al., 2003. Trisubstituted acridine derivatives as potent and selective telomerase inhibitors. J. Med. Chem. 46, 4463–4476.

Harrison, R.J., Gowan, S.M., Kelland, L.R., Neidle, S., 1999. Human telomerase inhibition by substituted acridine derivatives. Bioorg. Med. Chem. Lett. 9, 2463–2468.

Huang, H.S., Huang, K.F., Li, C.L., Huang, Y.Y., Chiang, Y.H., Huang, F.C., et al., 2008. Synthesis, human telomerase inhibition and anti-proliferative studies of a series of 2,7-bis-substituted amido-anthraquinone derivatives. Bioorg. Med. Chem. 16, 6976–6986.

Lane, A.N., Chaires, J.B., Gray, R.D., Trent, J.O., 2008. Stability and kinetics of G-quadruplex structures. Nucleic Acids Res. 36, 5482–5515.

Laronze-Cochard, M., Kim, Y.M., Brassart, B., Riou, J.-F., Laronze, J.-Y., Sapi, J., 2009. Synthesis and biological evaluation of novel 4,5-bis(dialkylaminoalkyl)-substituted acridines as potent telomeric G-quadruplex ligands. Eur. J. Med. Chem. 44, 3880–3888.

Leonetti, C., Amodei, S., D'Angelo, C., Rizzo, A., Benassi, B., Antonelli, A., et al., 2004. Biological activity of the G-quadruplex ligand RHPS4 (3,11-difluoro-6,8,13-trimethyl-8H-quino[4,3,2-kl]acridinium methosulfate) is associated with telomere capping alteration. Mol. Pharmacol. 66, 1138–1146.

Liu, Y., Wilson, W.D., 2010. Quantitative analysis of small molecule–nucleic acid interactions with a biosensor surface and surface plasmon resonance detection. Methods Mol. Biol. 613, 1–23.

Lombardo, C.M., Martínez, I.S., Haider, S., Gabelica, V., De Pauw, E., Moses, J.E., et al., 2010. Structure-based design of selective high-affinity telomeric quadruplex-binding ligands. Chem. Commun. (Camb.) 46, 9116–9118.

Manet, I., Manoli, F., Zambelli, B., Andreano, G., Masi, A., Cellai, L., et al., 2011. Affinity of the anthracycline antitumor drugs Doxorubicin and Sabarubicin for human telomeric G-quadruplex structures. Phys. Chem. Chem. Phys. 13, 540–551.

Martins, C., Gunaratnam, M., Stuart, J., Makwana, V., Greciano, O., Reszka, A.P., et al., 2007. Structure-based design of benzylamino-acridine compounds as G-quadruplex DNA telomere targeting agents. Bioorg. Med. Chem. Lett. 17, 2293–2298.

Mergny, J.-L., Phan, A.T., Lacroix, L., 1998. Following G-quartet formation by UV-spectroscopy. FEBS Lett. 435, 74–78.

Mergny, J.-L., Maurizot, J.-C., 2001. Fluorescence resonance energy transfer as a probe for G-quartet formation by a telomeric repeat. ChemBioChem 2, 124–132.

Monchaud, D., Teulade-Fichou, M.-P., 2008. A hitchhiker's guide to G-quadruplex ligands. Org. Biomol. Chem. 6, 627–636.

Moore, M.J., Schultes, C.M., Cuesta, J., Cuenca, F., Gunaratnam, M., Tanious, F.A., et al., 2006. Trisubstituted acridines as G-quadruplex telomere targeting agents. Effects of extensions of the 3,6- and 9-side chains on quadruplex binding, telomerase activity, and cell proliferation. J. Med. Chem. 49, 582–599.

Neidle, S., Harrison, R.J., Reszka, A.P., Read, M.A., 2000. Structure–activity relationships among guanine–quadruplex telomerase inhibitors. Pharmacol. Therapeut. 85, 133–139.

Olsen, C.M., Gmeiner, W.H., Marky, L.A., 2006. Unfolding of G-quadruplexes: energetic, and ion and water contributions of G-quartet stacking. J. Phys. Chem. B 110, 6962–6969.

Perry, P.J., Gowan, S.M., Reszka, A.P., Polucci, P., Jenkins, T.C., Kelland, L.R., et al., 1998. 1,4- and 2,6-disubstituted amidoanthracene-9,10-dione derivatives as inhibitors of human telomerase. J. Med. Chem. 41, 3253–3260.

Perry, P.J., Reszka, A.P., Wood, A.A., Read, M.A., Gowan, S.M., Dosanjh, H.S., et al., 1998. Human telomerase inhibition by regioisomeric disubstituted amidoanthracene-9,10-diones. J. Med. Chem. 41, 4873–4884.

Rachwal, P.A., Fox, K.R., 2007. Quadruplex melting. Methods 43, 291–301.

Read, M., Harrison, R.J., Romagnoli, B., Tanious, F.A., Gowan, S.H., Reszka, A.P., et al., 2001. Structure-based design of selective and potent G quadruplex-mediated telomerase inhibitors. Proc. Natl. Acad. Sci. USA 98, 4844–4849.

Read, M.A., Neidle, S., 2000. Structural characterization of a guanine–quadruplex ligand complex. Biochemistry 39, 13422–13432.

Read, M.A., Wood, A.A., Harrison, J.R., Gowan, S.M., Kelland, L.R., Dosanjh, H.S., et al., 1999. Molecular modeling studies on G-quadruplex complexes of telomerase inhibitors: structure–activity relationships. J. Med. Chem. 42, 4538–4546.

Ren, J., Chaires, J.B., 1999. Sequence and structural selectivity of nucleic acid binding ligands. Biochemistry 38, 16067–16075.

Rosu, F., De Pauw, E., Gabelica, V., 2008. Electrospray mass spectrometry to study drug–nucleic acids interactions. Biochimie 90, 1074–1087.

Schultes, C.M., Guyen, B., Cuesta, J., Neidle, S., 2004. Synthesis, biophysical and biological evaluation of 3,6-bis-amidoacridines with extended 9-anilino substituents as potent G-quadruplex-binding telomerase inhibitors. Bioorg. Med. Chem. Lett. 14, 4347–4351.

Sparapani, S., Haider, S.M., Doria, F., Gunaratnam, M., Neidle, S., 2010. Rational design of acridine-based ligands with selectivity for human telomeric quadruplexes. J. Amer. Chem. Soc. 132, 12263–12272.

Sun, D., Hurley, L.H., 2010. Biochemical techniques for the characterization of G-quadruplex structures: EMSA, DMS footprinting, and DNA polymerase stop assay. Methods Mol. Biol. 608, 65–79.

Sun, D., Thompson, B., Cathers, B.E., Salazar, M., Kerwin, S.M., Trent, J.O., et al., 1997. Inhibition of human telomerase by a G-quadruplex-interactive compound. J. Med. Chem. 40, 2113–2116.

Tan, J.H., Gu, L.Q., Wu, J.Y., 2008. Design of selective G-quadruplex ligands as potential anticancer agents. Mini Rev. Med. Chem. 8, 1163–1178.

Wang, Y., Patel, D.J., 1993. Solution structure of the human telomeric repeat $d[AG_3(T_2AG_3)_3]$ G-tetraplex. Structure 1, 263–282.

White, E.W., Tanious, F., Ismail, M.A., Reszka, A.P., Neidle, S., Boykin, D.W., et al., 2007. Structure-specific recognition of quadruplex DNA by organic cations: influence of shape, substituents and charge. Biophys. Chem. 126, 140–153.

Yang, D., Okamoto, K., 2010. Structural insights into G-quadruplexes: towards new anticancer drugs. Future Med. Chem. 2, 619–646.

Zagotto, G., Sissi, C., Lucatello, L., Pivetta, C., Cadamuro, S.A., Fox, K.R., et al., 2008. Aminoacyl–anthraquinone conjugates as telomerase inhibitors: synthesis, biophysical and biological evaluation. J. Med. Chem. 51, 5566–5574.

Zagotto, G., Sissi, C., Moro, S., Ben, D.D., Parkinson, G.N., Fox, K.R., et al., 2008. Amide bond direction modulates G-quadruplex recognition and telomerase inhibition by 2,6 and 2,7 bis-substituted anthracenedione derivatives. Bioorg. Med. Chem. 16, 354–361.

Zahler, A.M., Williamson, J.R., Cech, T.R., Prescott, D.M., 1991. Inhibition of telomerase by G-quartet DNA structures. Nature 350, 718–720.

Zambre, V.P., Murumkar, P.R., Giridhar, R., Yadav, M.R., 2010. Development of highly predictive 3D-QSAR CoMSIA models for anthraquinone and acridone derivatives as telomerase inhibitors targeting G-quadruplex DNA telomere. J. Mol. Graph. Model. 29, 229–239.

5

Telomeric Quadruplex Ligands II: Polycyclic and Non-fused Ring Compounds

Introduction

A large and diverse number of polycyclic and non-fused ring compounds have been investigated as telomeric quadruplex ligands. This chapter focuses on some of the more significant compound types, especially those that show potent quadruplex recognition and selectivity, although remarkably little structural information is as yet available on their complexes. Several excellent reviews are referenced at the end of the chapter, and these provide detailed accounts of particular areas. Among these many compounds are a number with excellent quadruplex-binding properties, yet whose biology has not been investigated to date in any detail. It is to be hoped that exposure here will encourage further effort on some of the more promising candidates, old and new. One category of quadruplex-binding compounds, those containing metals, has not been discussed here, purely for reasons of space. This is a rapidly expanding field and has links to nano materials as well as to potential therapeutic applications. The interested reader is referred to several excellent recent reviews of this topic.

More Polycyclic Compounds

The tricyclic phenanthridine dye ethidium bromide (Figure 5–1a) is a long-established duplex DNA binding agent, and on account of its fluorescent properties is sometimes used as a reagent in competition displacement assays for duplex-binding ligands. However, its binding to quadruplex DNAs is weak (Ren & Chaires, 1999), which can be rationalized by the presence of the non-coplanar phenyl substituent, which would limit the ability of the phenanthridine moiety to stack effectively with a G-quartet. Two strategies have been used to improve on ethidium. One involves attaching a (bulky) side-chain with a highly basic amidinium group (Koeppel *et al.*, 2001)—this resulted in a significant improvement in quadruplex affinity, dependent on the position of the substituent. The derivative shown in Figure 5–1b has a FRET ΔT_m value of 9.7° compared to that for ethidium itself, of 0.6°. It is also a potent telomerase inhibitor, as well as being

Therapeutic Applications of Quadruplex Nucleic Acids. DOI: 10.1016/B978-0-12-375138-6.00005-4

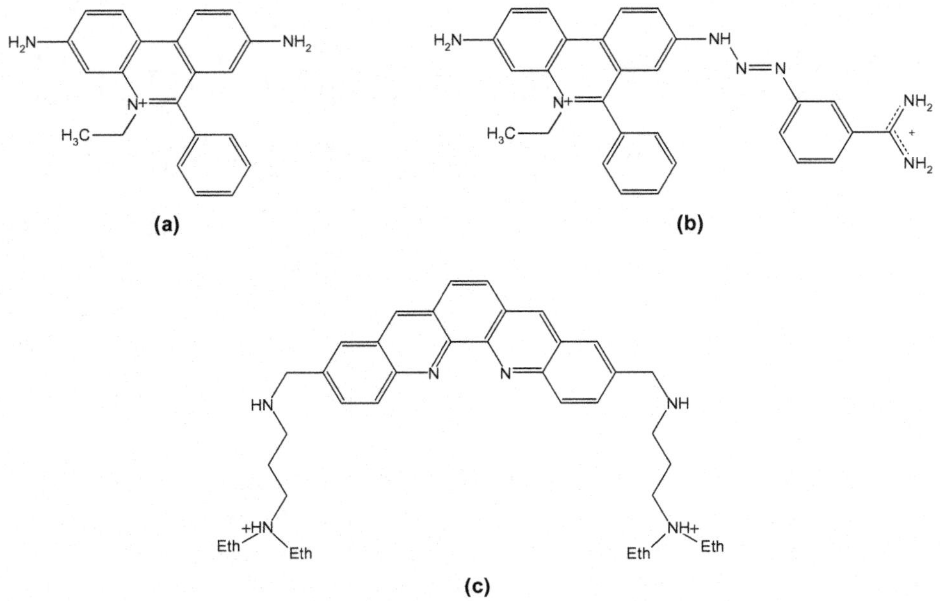

FIGURE 5–1 Structures of **(a)** ethidium, **(b)** an ethidium derivative, and **(c)** a phenanthroline derivative.

an anti-trypanocidal agent ("isomethamidium") used in veterinary medicine. However, these derivatives at least do not show significant selectivity for quadruplex over other nucleic acid structural types (Rosu *et al.*, 2003).

An alternative strategy has been to enlarge the size of the chromophore to, for example, a curve-shaped dibenzophenanthroline polycyclic core, remove the non-coplanar phenyl ring and add flexible alkylamino side-chains (Riou *et al.*, 2002). This results in markedly more potent compounds, such as the dibenzo[b,j]phenanthroline derivative shown in Figure 5–1c (Mergny *et al.*, 2001), which has a FRET ΔT_m value of 19.7° and a $^{tel}IC_{50}$ value of 0.028 μM (albeit from an early TRAP assay method). The majority of subsequent polycyclic ligands have continued on this approach of having an extended-size polycyclic as the core motif, to maximize overlap with G-quartets. Examples abound, including many flavors of quinolone and quindoline derivatives. Addition of two rather than a single side-chain, not unexpectedly, results in enhanced quadruplex binding and stabilization that correlates with telomerase inhibitory activity, as shown, for example, by a derivative of 10*H*-indolo[3,2-*b*]-quinoline (Figure 5–2a), which has a FRET ΔT_m value of ~15° at a 1 μM ligand concentration (Guyen *et al.*, 2004). Introduction of a formal positive charge on the pyridyl ring nitrogen atom was found to further improve quadruplex affinity (Lu *et al.*, 2008). An even more marked stabilization effect occurs on introducing two phenyl-ethoxyamino carboxamide substituents to the phenanthroline core (Wang *et al.*, 2011), with some compounds in this series having FRET ΔT_m values of >30° at a 1 μM ligand concentration together with high selectivity and significant telomerase inhibitory activity in the TRAP-LIG assay (see Chapter 1).

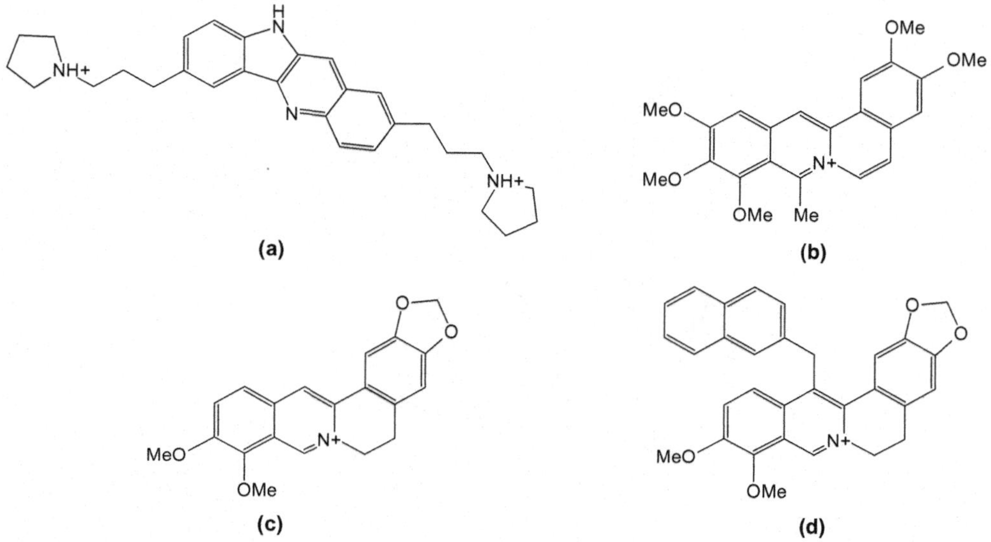

FIGURE 5–2 Structures of G-quadruplex binding quindoline and quinolone derivatives. See text for details.

The plant alkaloid berberine (Figure 5–2b) has modest quadruplex stabilizing ability (Franceschin *et al.*, 2006), probably because of the barrier to effective G-quartet stacking imposed by the non-planar central ring; thus the aromatized analog coralyne (Figure 5–2c) is much more effective, albeit with low selectivity, with a ΔT_m value over 20° higher than that of berberine. The addition of a naphthyl substituent (Figure 5–2d: Gornall *et al.*, 2010) results in superior affinity and markedly enhanced selectivity, with no binding to duplex DNA, as assessed by a mass spectrometric technique. It is plausible that the naphthyl substituent can reside in a quadruplex groove (Figure 5–3), analogous to the aniline group in the BRACO-19 complex with a human telomeric quadruplex (Chapter 4).

Porphyrins would appear to be ideal G-quadruplex ligands, with their extended conjugation available for optimal stacking with G-quartets. This was first demonstrated with the tetra-N-methylpyridyl porphyrin derivative, commonly known as TMPyP4 (Figure 5–4a), which was shown to bind to the intermolecular quadruplex formed from $d(T_4G_4)$ with ~2-fold greater affinity than to duplex DNA (Anantha, Azam & Sheardy, 1998). A large number of subsequent studies have used TMPyP4 as a starting point to probe for quadruplex structure, although the molecular details and even the stoichiometry of the interactions between TMPyP4 and human telomeric quadruplexes remain unclear. The ability of TMPyP4 to bind quadruplex and other forms of DNA with equal felicity has been confirmed by several studies (see, for example, Ren & Chaires, 1999; Wang *et al.*, 2006). TMPyP4 unsurprisingly produces a wide range of effects in biological systems as a consequence of its multi-targeting—it has anti-tumor activity in xenograft models (Grand *et al.*, 2002) and affects the transcription of a large number of genes including those in which there are quadruplex-forming sequences in their promoters (see Chapter 7). TMPyP4 and many of its analogs are potent telomerase inhibitors (Shi *et al.*, 2001), with

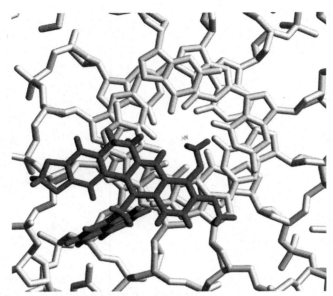

FIGURE 5–3 Details of a molecular model simulating the binding of compound d in Figure 5.2 to a human telomeric quadruplex. The ligand is shown in shaded stick bonds. Note the almost 45° orientation of the naphthalene group with respect to the G-quartet plane, which enables it to lie in a quadruplex groove.

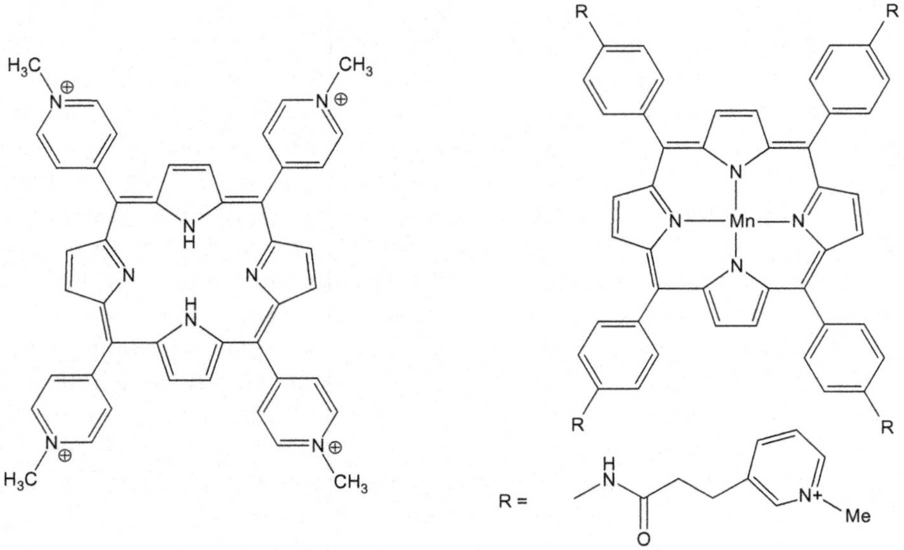

(a)

(b)

FIGURE 5–4 Structure of **(a)** the TMPyP4 molecule and **(b)** a pentacationic manganese porphyrin molecule (Dixon et al., 2007).

a $^{tel}IC_{50}$ value of 6.5 μM in a direct telomerase assay (see Chapter 1 for further details). It is possible to devize TMPyP4 analogs with improved quadruplex selectivity. Thus the pentacationic manganese porphyrin compound (Figure 5–4b) has 10^4-fold selectivity for the human telomeric quadruplex sequence compared to its duplex affinity constant, as measured by surface plasmon resonance methods (Dixon *et al.*, 2007). It is likely that this effect is a consequence of the long side-arms rather than the presence of the manganese ion since metal ions in porphyrin centers do not normally affect quadruplex affinity or selectivity. In general increasing the size of the peripheral substituents on the porphyrin ring, or modifying the ring system itself, for example to a phthalocyanine system, enhances selectivity, but at the price of molecular weight and size (Alzeer & Luedtke, 2010).

A study of TMPyP4 binding to human telomeric quadruplexes in molecular crowding conditions has concluded that the stoichiometry can be a maximum of 4:1, with the parallel quadruplex topology being preferred, although not in dilute solution, when an anti-parallel form is dominant (Martino *et al.*, 2009). The crystal structure of TMPyP4 complexed with a human telomeric bimolecular quadruplex having a parallel topology (Parkinson, Ghosh & Neidle, 2007) shows porphyrin molecules externally stacking onto the bases of the TTA propeller loops, as well as one porphyrin stacked between two quadruplexes. However, this TMPyP4 molecule is not stacked between G-quartets but rather between A·T base pairs formed from the reorganization of the loops (Figure 5–5). It is unlikely that direct stacking onto the ends of a quadruplex core is possible since the four N-methyl pyridyl groups in TMPyP4 are inherently non-planar with respect to the

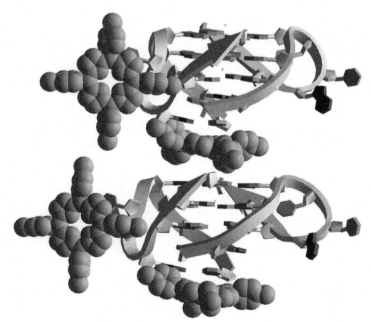

FIGURE 5–5 Representation of the crystal structure of a complex between the TMPyP4 molecule and an 11-mer bimolecular telomeric quadruplex (Parkinson, Ghosh & Neidle, 2007: PDB id 2HRI).

porphyrin core and cannot stack with G-quartets at the normal 3.4Å separation—this increased separation is seen in the only other structure available for a TMPyP4–quadruplex complex, in the NMR structure with the *c-myc* promoter quadruplex (Phan *et al.*, 2005) where the porphyrin and associated G-quartet are 4.5Å apart. Instead the inside edges of the attached N-methyl pyridyl rings are in close 3.4Å contact with the G-quartet face. These structural features are consistent with the stoichiometry and binding assignments inferred from a detailed biophysical study (Wei *et al.*, 2009), in the absence of further crystal and NMR structures.

The naphthalene diimide (ND) platform (Figure 5–6a) was originally conceived of as a chromophore that would cover a significant proportion of G-quartet 2-D space (Cuenca *et al.*, 2008). This was supplemented with three or four side-chain substituents having the "standard" alkylamino functionality. Several members of the resulting small library of tri- and tetra-substituted ND compounds showed exceptionally high quadruplex stabilizing properties, with ΔT_m values of up to 30° for a human telomeric quadruplex even at a ligand concentration of 0.5 μM, and some selectivity for quadruplex over duplex DNA. The most active compounds in the series are potent telomerase inhibitors and are highly active antiproliferative agents, with IC_{50} values in a range of cancer cell lines averaging ~100 nM. These molecules are rapidly transported into cell nuclei (in spite of their size and high cationic charge) and their intense fluorescence has enabled uptake to be readily visualized. Molecular modeling (described in Chapter 9) suggested replacement of a terminal N-diethyl, pyrrolidino or piperidino group with an N-methyl piperazino group (Figure 5–6b), on a slightly longer side-chain, which has resulted in some enhancement of quadruplex vs. duplex selectivity (Hampel *et al.*, 2010). Contrary to expectation

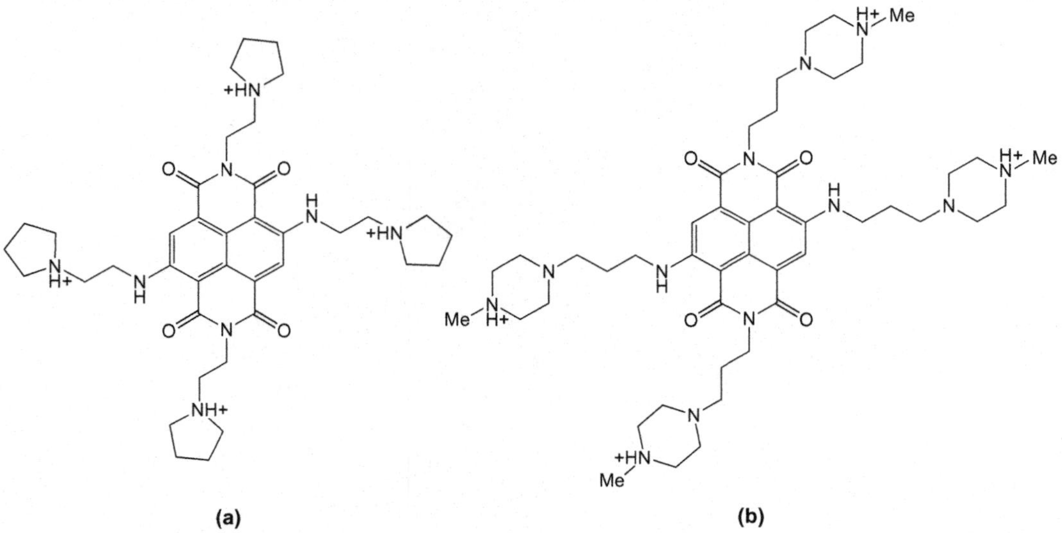

(a) (b)

FIGURE 5–6 Two naphthalene diimide compounds, **(a)** with pyrrolidino end-groups on the side-chains (Cuenca *et al.*, 2008), and **(b)** with N-methyl piperazine end-groups (Hampel *et al.*, 2010).

the side-chain cationic groups have few direct interactions with quadruplex phosphate groups in the grooves. Rather they interact with water molecules, some of which mediate between the cationic groups and hydrogen-bond acceptors/donors on the deoxyribose sugar backbone and to nucleobase edges. These structures also suggest that yet longer side-chains could make enhanced contacts, but to date this is conjecture since ligands with these features have not been successfully co-crystallized with a quadruplex.

Two crystal structures of ND complexes with human telomeric quadruplexes and a ND ligand have been determined, one of an intramolecular quadruplex (Figure 5–7a) and the other with a bimolecular quadruplex (Parkinson, Cuenca & Neidle, 2008). A number of structural changes in loop geometries compared to the native structure, and to the BRACO-19 complex, are apparent, which probably reflects the differing binding requirements of individual ligands. The extent of overlap of the naphthalene diimide core with G-quartets (Figure 5–7b) is not extensive, in accord with the experimental findings that the four side-chains and their cationic charges are the major contributors to the binding. The reduction in ΔT_m value on reducing the number of cationic side-chains from four to three is ~6–7°, and the reduction on substituting four charged end-groups for uncharged (but bulky) morpholino groups is ~15°. A further reduction, of ~7–10°, has been observed for those compounds having three morpholino side-chains. Thus the contribution of the naphthalene diimide platform to the overall binding is low, and it is really acting as a delivery vehicle for the side-chains. Total removal abolishes binding completely.

The ND compounds have quasi four-fold symmetry, which should, if the side-chains have sufficient reach, exploit the four-fold symmetry of the (parallel form) of the human

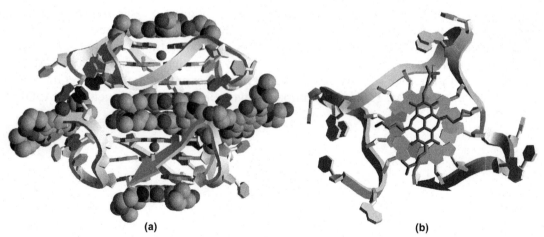

(a)   (b)

FIGURE 5–7 (a) Representation of the crystal structure of a naphthalene diimide (ND) ligand complexed with a human intramolecular quadruplex formed from d(TAGGG(TTAGGG)₃). The structure has two quadruplexes stacked end to end, with two ND ligands sandwiched between them, shown in space-filling mode. Other ND ligands are shown stacked onto the bases of the TTA loops (Parkinson, Cuenca & Neidle, 2008: PDB id 3CDM). **(b)** Showing a view of the same structure onto the G-terminal quartet of one quadruplex, with the ND ligand shown in stick representation.

telomeric quadruplex, with its four grooves. The binding data suggest that the optimal compounds do this rather well. A counter-intuitive approach is to evaluate molecules with three-fold symmetry, which also have extensive potential for π–π overlap with a G-quartet surface. Star-shaped series of compounds have been devised, for example based on the triazatrinaphthylene core and with three standard aminoalkyl side-chains (Bertrand *et al.*, 2011: Figure 5–8). These can be thought of as acridines extended in two further directions, although the concept of three side-chains was initially successful with tri-substituted mono-acridines (see Chapter 4). As with many other quadruplex ligands, the extent of binding has been found to be dependent on charge and side-chain length. The optimal ligands in this and other star-shaped series do have high quadruplex affinity, although their ability to discriminate between different quadruplexes is limited.

Telomestatin—the Archetypal Macrocyclic Quadruplex Ligand

This unique cyclic natural product containing a pentaoxazole ring system plus a thiazoline ring was first isolated from *Streptomyces anulatus* (Shin-ya *et al.*, 2001) and has one asymmetric center. Almost all studies up to very recently have been performed with the racemate. Telomestatin (Figure 5–9) has exceptional quadruplex-binding and telomerase inhibitory properties. An initial value for the telomerase activity gave a $^{tel}IC_{50}$ value of 0.005 μM using a TRAP assay, but has been revised on the basis of a direct telomerase assay (De Cian *et al.*, 2007) to possibly 1500-fold less. However, this downward revision affects all G-quadruplex-binding telomerase inhibitors and telomestatin still remains one of the most potent G-quadruplex ligands studied to date. No experimental structural information is available on a quadruplex–telomestatin complex, but molecular modeling studies indicate that the cyclic, almost planar telomestatin molecule is of comparable size to a G-quartet and therefore they can have excellent overlap (Figure 5–10), in accordance with the high-affinity constant determined experimentally (Kim *et al.*, 2002; Luedtke, 2009) of ~30 nM. This value compares favorably with ligands such as BRACO-19

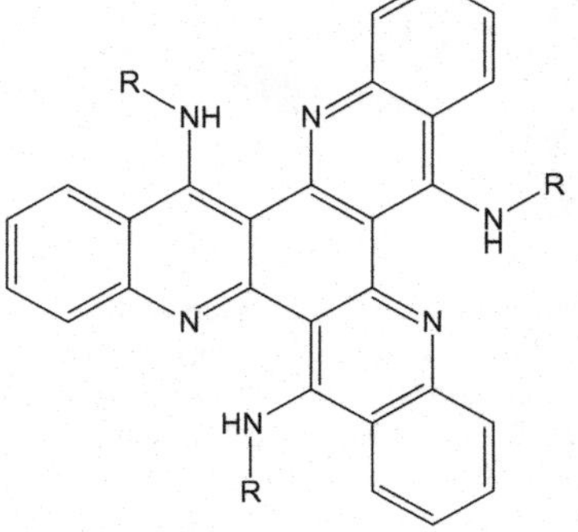

FIGURE 5–8 The structure of a substituted triazatrinaphthylene derivative. R = -(CH$_2$)$_3$-NH$_3$$^+$ (Bertrand *et al.*, 2011).

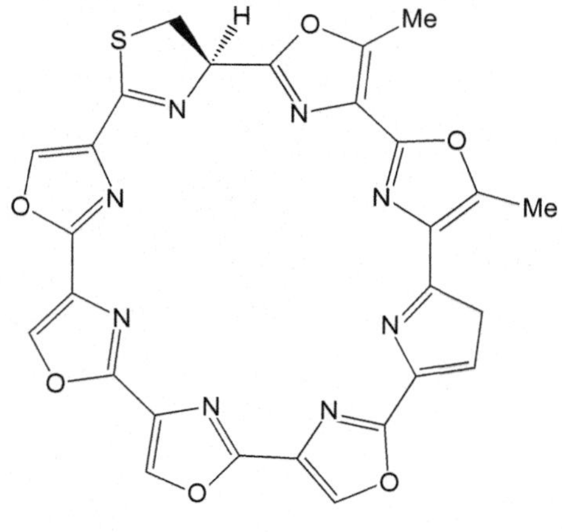

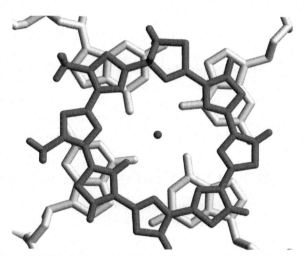

FIGURE 5–9 The structure of telomestatin, showing the natural (*R*) absolute configuration.

FIGURE 5–10 A molecular model illustrating a plausible binding mode of telomestatin to a G-quartet. The ligand is shown in shaded stick representation, and the telomeric quadruplex used here is the human 22-mer crystal structure (PDB id 1KF1).

(~500 nM) and RHPS4 (~100 nM), and unsurprisingly the behavior of telomestatin has been examined in a number of biological studies (see Chapters 6 and 7). These have demonstrated its potency as a telomere targeting agent and telomerase inhibitor. Further development of telomestatin as a potential therapeutic agent has been hampered until recently by its poor aqueous solubility and lack of supply. Total syntheses of telomestatin as the natural (*R*) stereoisomer have been reported (Doi *et al.*, 2006, and more recently by Linder *et al.*, 2011). These achievements will also enable derivatives of telomestatin to be synthesized in the future in order to enhance its pharmacological properties, especially

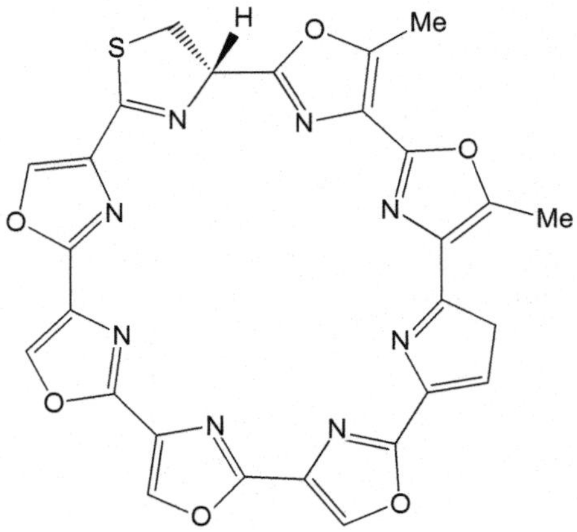

FIGURE 5–11 The structure of (*S*)-telomestatin.

its aqueous solubility, perhaps by addition of an appropriately placed small cationic side-chain.

Synthesis of the unnatural (*S*) isomer of telomestatin (Figure 5–11) has recently been achieved (Doi *et al.*, 2011). This compound appears to be significantly more potent as a quadruplex stabilizer than the natural (*R*) isomer, with an enhancement in ΔT_m of 6° on binding to a human telomeric quadruplex, and as a telomerase inhibitor, with a four-fold improvement in $^{tel}IC_{50}$. The origin of these differences is unclear at present.

Other Non-polycyclic Quadruplex Ligands

A number of macrocyclic mimetics of telomestatin have been described in the literature, many of which are synthetically more accessible than telomestatin. They sometimes have one or more of the ubiquitous pendant alkylamino side-chains, which enhance quadruplex affinity and solubility. They range from tri-oxazoles (Jantos *et al.*, 2006; Rzuczek *et al.*, 2010) to hexa-oxazoles (Barbierei *et al.*, 2007; Tera *et al.*, 2008). In order to achieve maximum shape and size compatibility with telomestatin, several distinct bridging strategies have been used to link oxazole groups (Figure 5–12a). Thus, amide groups or simple peptide groups such as valine have been employed (Jantos *et al.*, 2006; Minhas *et al.*, 2006), as well as pyridyl groups (Rzuczek *et al.*, 2010) or even all-carbon acyclic linkers (Satyanaranayana *et al.*, 2006). The circular shape of telomestatin itself is distorted slightly in the analog with a pyridyl and a phenyl ring forming part of the macrocycle (Figure 5–12b: Rzuczek *et al.*, 2010). This compound thus has a slightly oval shape and is not fully coplanar, factors that reduce the overlap with a G-quartet compared to telomestatin itself (Figure 5–13). However, the presence of the solubilizing "standard" alkylamino side-chain is likely to aid quadruplex binding,

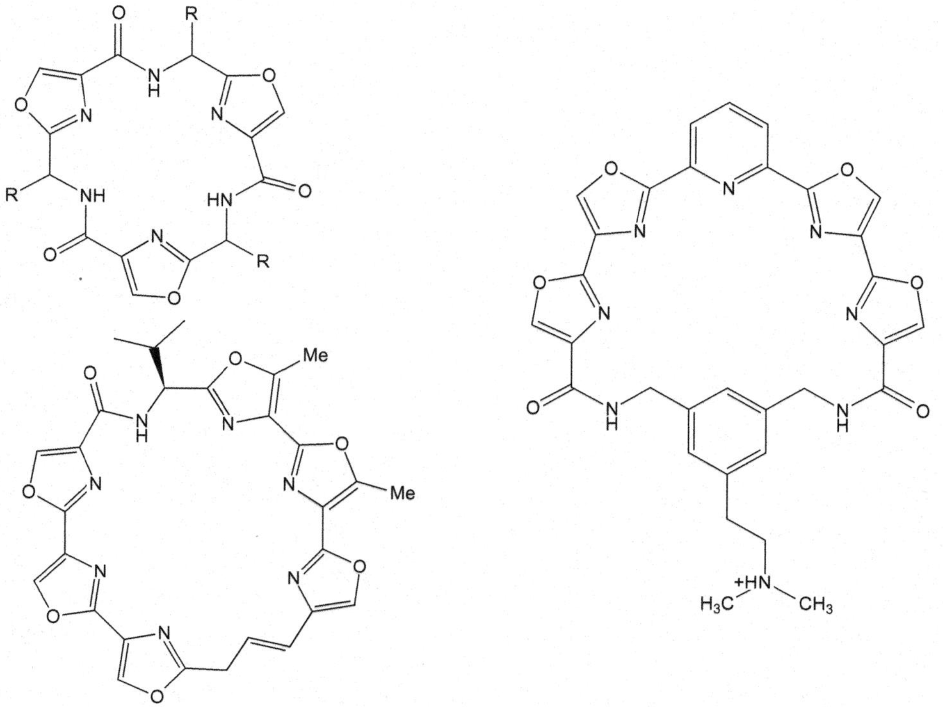

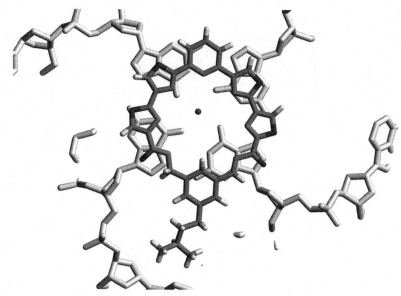

FIGURE 5–12 (a) Structures of telomestatin mimetics: a cyclic peptidic trisoxazole (Jantos *et al.*, 2006) and a cyclic hexoxazole (Minhas *et al.*, 2006). **(b)** A cyclic hexoxazole containing pyridyl and phenyl rings as part of the macrocycle (Rzuczek *et al.*, 2010).

FIGURE 5–13 A molecular model showing a plausible binding mode of the pyridyl-phenyl hexoxazole (Rzuczek *et al.*, 2010) onto a G-quartet. The ligand is shown in shaded stick representation, and the telomeric quadruplex used here is the human 22-mer crystal structure (PDB id 1KF1). The aminoalkyl side-chain of the ligand is shown lying in the groove of this quadruplex.

as indicated by a ΔT_m value of 20.5° with a human telomeric quadruplex. Its ability to stabilize an mRNA quadruplex from aurora kinase is impressive, with a ΔT_m of 37° and no discernible affinity with either duplex DNA or RNA. This synthetically accessible compound is also highly active in both various cancer cell lines and in an *in vivo* tumor xenograft model. Although it remains to be demonstrated that these biological effects are a consequence of (DNA or RNA) quadruplex targeting, the data do indicate that telomestatin mimicry is able to generate novel lead compounds with potential for further evaluation in humans.

A number of other chemically diverse compound classes characterized by linked (mostly) monocyclic rings have quadruplex-binding ability. Many utilize amide linkages, and some have been conceived as open-chain precursors of telomestatin-like molecules and a number have significant activity in their own right. The duplex DNA binding natural product antibiotic distamycin A (Figure 5–14) comprises one such molecule, with N-methylpyrrole rings linked by amide groups. It has been reported to bind to several quadruplex types, albeit with low selectivity and affinity. Thus its ΔT_m to a duplex DNA sequence is 20.5°, whereas it barely stabilizes the intramolecular telomeric quadruplex, with a ΔT_m of 3.5° (Moore *et al.*, 2006). The mode of binding to quadruplexes has recently been clarified, following an NMR structure determination for a 4:1 distamycin analog complex with the tetramolecular quadruplex formed by four strands of d(TGGGGT) (Cosconati *et al.*, 2010). This is an unusual structure since it is the first quadruplex–ligand complex structure to be determined in which G-quartet stacking is not a major factor. Instead (Figure 5–15a, b: PDB id 2KVY), the crescent-shaped ligand binds as two sets of dimers across the 3′ end of the quadruplex so that each dimer is embedded into a groove. This arrangement is reminiscent of the minor groove binding of distamycin and many of its analogs to duplex DNA. Attempts to move away from groove binding back to the G-quartet surface yet retain the synthetically accessible features of distamycin-like molecules have been reported by Rahman *et al.* (2009), and are discussed further in Chapter 9.

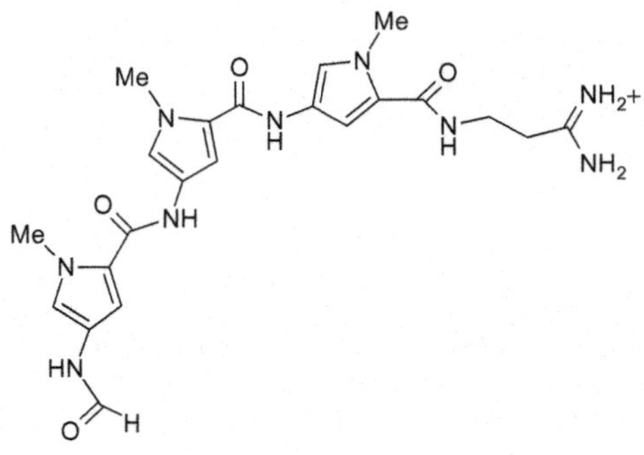

FIGURE 5–14 The structure of distamycin A.

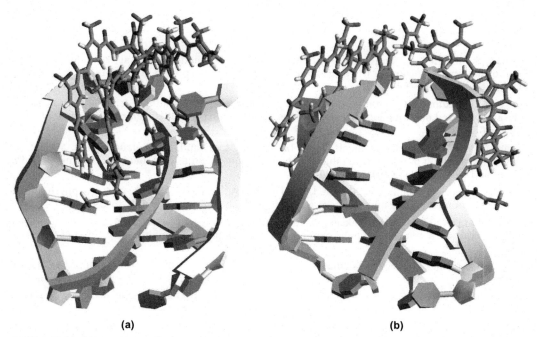

(a)   **(b)**

FIGURE 5–15 (a), (b) Two views of one structure taken from the ensemble of NMR structures for the distamycin derivative–d(TGGGGT) complex.

Recommended Reviews

Georgiades, S.N., Abd Karim, N.H., Suntharalingam, K., Vilar, R., 2010. Interaction of metal complexes with G-quadruplex DNA. Angew. Chem. Int. Ed. Engl. 49, 4020–4034.

Monchaud, D., Teulade-Fichou, M.-P., 2008. A hitchhiker's guide to G-quadruplex ligands. Org. Biomol. Chem. 6, 627–636.

Nielsen, M.C., Ulven, T., 2010. Macrocyclic G-quadruplex ligands. Curr. Med. Chem. 17, 3438–3448.

Ralph, S.F., 2011. Quadruplex DNA: a promising drug target for the medicinal inorganic chemist. Curr. Topics Med. Chem. 11, 572–590.

Tan, J.H., Gu, L.Q., Wu, J.Y., 2008. Design of selective G-quadruplex ligands as potential anticancer agents. Mini Rev. Med. Chem. 8, 1163–1178.

References

Alzeer, J., Luedtke, N.W., 2010. pH-mediated fluorescence and G-quadruplex binding of amido phthalocyanines. Biochemistry 49, 4339–4348.

Anantha, N.V., Azam, M., Sheardy, R.D., 1998. Porphyrin binding to quadruplexed T_4G_4. Biochemistry 37, 2709–2714.

Barbieri, C.M., Srinivasan, A.R., Rzuczek, S.G., Rice, J.E., LaVoie, E.J., Pilch, D.S., 2007. Defining the mode, energetics and specificity with which a macrocyclic hexaoxazole binds to human telomeric G-quadruplex DNA. Nucleic Acids Res. 35, 3272–3286.

Bertrand, H., Granzhan, A., Monchaud, D., Saettel, N., Guillot, R., Clifford, S., et al., 2011. Recognition of G-quadruplex DNA by triangular star-shaped compounds: with or without side chains? Chemistry 17, 4529–4539.

Cosconati, S., Marinelli, L., Trotta., R., Virno, A., De Tito, S., Romagnoli, R., et al., 2010. Structural and conformational requisites in DNA quadruplex groove binding: another piece to the puzzle. J. Amer. Chem. Soc. 132, 6425–6433.

Cuenca, F., Greciano, O., Gunaratnam, M., Haider, S., Munnur, D., Nanjunda, R., et al., 2008. Tri- and tetra-substituted naphthalene diimides as potent G-quadruplex ligands. Bioorg. Med. Chem. Lett. 18, 1668–1673.

De Cian, A., Cristofai, G., Reichenbach, P., De Lemos, E., Monchaud, D., Teulade-Fichou, M.-P., et al., 2007. Reevaluation of telomerase inhibition by quadruplex ligands and their mechanisms of action. Proc. Natl. Acad. Sci. USA 104, 17347–17352.

Dixon, I.M., Lopez, F., Tejera, A.M., Estève, J.-P., Blasco, M.A., Pratviel, G., et al., 2007. Porphyrin derivatives for telomere binding and telomerase inhibition. J. Amer. Chem. Soc. 129, 1502–1503.

Doi, T., Shibata, K., Yoshida, M., Takagi, M., Tera, M., Nagasawa, K., et al., 2011. (S)-Stereoisomer of telomestatin as a potent G-quadruplex binder and telomerase inhibitor. Org. Biomol. Chem. 9, 387–393.

Doi, T., Yoshida, M., Shin-ya, K., Takahashi, T., 2006. Total synthesis of (R)-telomestatin. Org. Lett. 8, 4165–4167.

Franceschin, M., Rossetti, L., D'Ambrosio, A., Schirripa, S., Bianco, A., Ortaggi, G., et al., 2006. Natural and synthetic G-quadruplex interactive berberine derivatives. Bioorg. Med. Chem. Lett. 16, 1707–1711.

Gornall, K.C., Samosorn, S., Tanwirat, B., Suksamrarn, A., Bremner, J.B., Kelso, M.J., et al., 2010. A mass spectrometric investigation of novel quadruplex DNA-selective berberine derivatives. Chem. Commun. (Camb.) 46, 6602–6604.

Grand, C.L., Han, H., Muñoz, R.M., Weitman, S., Von Hoff, D.D., Hurley, L.H., et al., 2002. The cationic porphyrin TMPyP4 down-regulates c-MYC and human telomerase reverse transcriptase expression and inhibits tumor growth in vivo. Mol. Cancer Ther. 1, 565–573.

Guyen, B., Schultes, C.M., Hazel, P., Mann, J., Neidle, S., 2004. Synthesis and evaluation of analogues of 10H-indolo[3,2-b]quinoline as G-quadruplex stabilising ligands and potential inhibitors of the enzyme telomerase. Org. Biomol. Chem. 2, 981–988.

Hampel, S.M., Sidibe, A., Gunaratnam, M., Riou, J.-F., Neidle, S., 2010. Tetrasubstituted naphthalene diimide ligands with selectivity for telomeric G-quadruplexes and cancer cells. Bioorg. Med. Chem. Lett. 20, 6459–6463.

Jantos, K., Rodriguez, R., Ladame, S., Shirude, P.S., Balasubramanian, S., 2006. Oxazole-based peptide macrocycles: a new class of G-quadruplex binding ligands. J. Amer. Chem. Soc. 128, 13662–13663.

Kim, M.-Y., Vankayalapati, H., Shin-ya, K., Wierzba, K., Hurley, L.H., 2002. Telomestatin, a potent telomerase inhibitor that interacts quite specifically with the human telomeric intramolecular G-quadruplex. J. Amer. Chem. Soc. 124, 2098–2099.

Koeppel, F., Riou, J.-F., Laoui, A., Mailliet, P., Arimondo, P.B., Labit, D., et al., 2001. Ethidium derivatives bind to G-quartets, inhibit telomerase and act as fluorescent probes for quadruplexes. Nucleic Acids Res. 29, 1087–1096.

Linder, J., Garner, T.P., Williams, H.E.L., Searle, M.S., Moody, C.J., 2011. Telomestatin: formal total synthesis and cation-mediated interaction of its seco-derivatives with G-quadruplexes. J. Amer. Chem. Soc. 133, 1044–1051.

Lu, Y.-J., Ou, T.-M., Tan, J.-H., Hou, J.-Q., Shao, W.-Y., Peng, D., et al., 2008. 5-N-Methylated quindoline derivatives as telomeric G-quadruplex stabilizing ligands: effects of 5-N positive charge on quadruplex binding affinity and cell proliferation. J. Med. Chem. 51, 6381–6392.

Luedtke, N.W., 2009. Targeting G-quadruplex DNA with small molecules. Chimia 63, 134–139.

Martino, L., Pagano, B., Fotticchia, I., Neidle, S., Giancola, C., 2009. Shedding light on the interaction between TMPyP4 and human telomeric quadruplexes. J. Phys. Chem. B 113, 11479–11486.

Mergny, J.-L., Lacroix, L., Teulado-Fichou, M.-P., Hounsou, C., Guittat, L., Hoarau, M., et al., 2001. Telomerase inhibitors based on quadruplex ligands selected by a fluorescence assay. Proc. Natl. Acad. Sci. USA 98, 3062–3067.

Minhas, G.S., Pilch, D.S., Kerrigan, J.L., LaVoie, E.J., Rice, J.E., 2006. Synthesis and G-quadruplex stabilizing properties of a series of oxazole-containing macrocycles. Bioorg. Med. Chem. Lett. 16, 3891–3895.

Moore, M.J., Cuenca, F., Searcey, M., Neidle, S., 2006. Synthesis of distamycin A polyamides targeting G-quadruplex DNA. Org. Biomol. Chem. 4, 3479–3488.

Parkinson, G.N., Cuenca, F., Neidle, S., 2008. Topology conservation and loop flexibility in quadruplex–drug recognition: crystal structures of inter- and intramolecular telomeric DNA quadruplex–drug complexes. J. Mol. Biol. 381, 1145–1156.

Parkinson, G.N., Ghosh, R., Neidle, S., 2007. Structural basis for binding of porphyrin to human telomeres. Biochemistry 46, 2390–2397.

Phan, A.T., Kuryavyi, V., Gaw, H.Y., Patel, D.J., 2005. Small-molecule interaction with a five-guanine-tract G-quadruplex structure from the human MYC promoter. Nature Chem. Biol. 1, 167–173.

Rahman, K.M., Reszka, A.P., Gunaratnam, M., Haider, S.M., Howard, P.W., Fox, K.R., et al., 2009. Biaryl polyamides as a new class of DNA quadruplex-binding ligands. Chem. Commun. (Camb.), 4097–4099.

Ren, J., Chaires, J.B., 1999. Sequence and structural selectivity of nucleic acid binding ligands. Biochemistry 38, 16067–16075.

Riou, J.-F., Guittat., L., Mailliet, P., Laoui, A., Renou, E., Petitgenet, O., et al., 2002. Cell senescence and telomere shortening induced by a new series of specific G-quadruplex DNA ligands. Proc. Natl. Acad. Sci. USA 99, 2672–2677.

Rosu, F., De Pauw, E., Guittat, L., Alberti, P., Lacroix, L., Mailliet, P., et al., 2003. Selective interaction of ethidium derivatives with quadruplexes: an equilibrium dialysis and electrospray ionization mass spectrometry analysis. Biochemistry 42, 10361–10371.

Rzuczek, S.G., Pilch, D.S., Liu, A., Liu, L., LaVoie, E.J., Rice, J.E., 2010. Macrocyclic pyridyl polyoxazoles: selective RNA and DNA G-quadruplex ligands as antitumor agents. J. Med. Chem. 53, 3632–3644.

Satyanarayana, M., Kim, Y.A., Rzuczek, S.G., Pilch, D.S., Liu, A.A., Liu, L.F., et al., 2010. Macrocyclic hexaoxazoles: influence of aminoalkyl substituents on RNA and DNA G-quadruplex stabilization and cytotoxicity. Bioorg. Med. Chem. Lett. 20, 3150–3154.

Shi, D.F., Wheelhouse, R.T., Sun, D., Hurley, L.H., 2001. Quadruplex-interactive agents as telomerase inhibitors: synthesis of porphyrins and structure–activity relationship for the inhibition of telomerase. J. Med. Chem. 44, 4509–4523.

Shin-ya, K., Wierzba, K., Matsuo, K., Ohtani, T., Yamada, Y., Furihata, K., et al., 2001. Telomestatin, a novel telomerase inhibitor from *Streptomyces anulatus*. J. Amer. Chem. Soc. 123, 1262–1263.

Tera, M., Ishizuka, H., Takagi, M., Suganuma, M., Shin-ya, K., Nagasawa, K., 2008. Macrocyclic hexaoxazoles as sequence- and mode-selective G-quadruplex binders. Angew. Chem. Int. Ed. Engl. 47, 5557–5560.

Wang, P., Ren, L., He, H., Liang, F., Zhou, X., Tan, Z., 2006. A phenol quaternary ammonium porphyrin as a potent telomerase inhibitor by selective interaction with quadruplex DNA. Chembiochem 7, 1155–1159.

Wang, L., Wen, Y., Liu, J., Zhou, J., Li, C., Wei, C., 2011. Promoting the formation and stabilization of human telomeric G-quadruplex DNA, inhibition of telomerase and cytotoxicity by phenanthroline derivatives. Org. Biomol. Chem. 9, 2648–2653.

Wei, C., Jia, G., Zhou, J., Han, G., Li, C., 2009. Evidence for the binding mode of porphyrins to G-quadruplex DNA. Phys. Chem. Chem. Phys. 11, 4025–4032.

6

The Biology and Pharmacology of Telomeric Quadruplex Ligands

Introduction

We have described in Chapter 1 the classic model of telomerase inhibition and that the consequences of telomere attrition involve a long time lag before they irreversibly affect cellular function and viability. This extended time lag was initially observed in telomerase inhibition experiments in cancer cell lines using direct inhibitors of hTERT, the catalytic domain of human telomerase, but it was soon realized that the time lag would be therapeutically challenging for human cancer treatment since it would allow tumors to continue growing during the lag time. On the other hand, experiments using G-quadruplex ligands from the outset showed very different behavior. In the case of BRACO-19 (Chapter 4) senescence was observed within 15 days after cells were first treated, and there was no evidence of concomitant telomere shortening after exposure for this length of time (Gowan *et al.*, 2002). Encouragingly, significant anti-tumor activity was observed in the A431 human vulval carcinoma xenograft model in combination with paclitaxel after just seven days. Telomere shortening of ~0.4 kb was observed (Burger *et al.*, 2005) after 15 days' treatment with BRACO-19 in the human uterine carcinoma cell line UXF1138L, which has very short telomeres (2.7 kb). This loss corresponds to ~34 nucleotides per round of cell division in these cells, which have a population doubling time of 30.5 hours. BRACO-19 is also active *in vivo* against the xenograft raised from this cell line, but on a much shorter time-scale, with responses within 4–5 days (see below). Analogous results have been obtained with several other ligands. Telomestatin produced potent telomerase inhibition, dose-dependent cytotoxicity and apoptosis in neuroblastoma cells following short-term (72 h) exposure; long-term subcytotoxic exposure led to telomere shortening as expected (Binz *et al.*, 2005). Telomere shortening was observed in the human leukemic lymphoma cell line U937 after 20 population doublings on treatment with telomestatin (Tauchi *et al.*, 2006). On the other hand, cell lines containing short telomeres at the outset would be expected to demonstrate classic behavior on a short time-scale, as was found with Barrett's-associated esophageal adenocarcinoma cells (Shammas *et al.*, 2004). These have short telomeres of mean length 2.4–3.3 kb and a 2,6-bis-piperidino amidoanthraquinone derivative has been shown to produce shortening and telomerase inhibition within the predicted timescale.

Therapeutic Applications of Quadruplex Nucleic Acids. DOI: 10.1016/B978-0-12-375138-6.00006-6

Taken together, data from these and a number of studies on telomere-targeted quadruplex ligands are strongly suggestive of more than one mechanism of action, involving short-term and long-term responses at the telomere (Gunaratnam *et al.*, 2007). This duality of action has been subsequently shown to be characteristic of the telomeric G-quadruplex ligand class as a whole (Riou, 2004; De Cian *et al.*, 2008), and the ligands are sometimes now termed Telomere Targeting Agents (TTAs) (Phatak & Burger, 2007; Neidle, 2010) to reflect this emphasis. The long-term effects can be attributed to the classic telomerase inhibition pathway of ligand-induced telomere erosion, followed by senescence and apoptosis. Cells (and tumors) with short telomeres are unsurprisingly more sensitive (and respond more rapidly) to loss of telomere function than those with much extended telomeres (Hemann *et al.*, 2001). For example, the response to the polycyclic acridine derivative RHPS4 of several MCF-7 human breast carcinoma cell lines transfected with a range of restriction fragment lengths has been found to correlate with telomere length (Cookson *et al.*, 2005). The consistent observations of *in vitro* and *in vivo* activity for G-quadruplex ligands within clinically useful timescales coupled with evidence of selective action at telomeres are encouraging signs that significant single-agent clinical utility may be achievable in the future with appropriate compounds.

G-quadruplex Ligands Induce DNA Damage Responses

The quadruplex-binding acridine ligands BRACO-19 and RHPS4 in common with telomestatin produce a number of short-term changes that are not associated with telomere shortening, such as induction of senescence, anaphase bridges and chromosomal end-to-end fusions (Incles *et al.*, 2004; Tahara *et al.*, 2006; Gunaratnam *et al.*, 2007). Typically there is induction of rapid replicative senescence in cancer cells. These changes at the telomere also trigger the activation of a series of DNA damage responses that follows DNA double-strand breaks. The responses involve in particular the ATR/ATM signaling pathways (Tauchi *et al.*, 2003; Pennarun *et al.*, 2008; Rizzo *et al.*, 2009; Zhou *et al.*, 2009), together with p16[INK4a] kinase and p53 pathways (Phatak *et al.*, 2007; Salvati *et al.*, 2007). The DNA damage response can be monitored by the appearance of characteristic DNA damage foci using an antibody to the damage-response protein marker γ-H2AX (Rodriquez *et al.*, 2008; Casagrande *et al.*, 2011). The resulting anti-tumor responses observed in animal models can be directly attributable to DNA damage (Salvati *et al.*, 2007).

These lethal cellular events are responses to the displacement of bound hPOT1 and other proteins from the single-stranded overhang (Gomez *et al.*, 2006a, b; Gunaratnam *et al.*, 2007; Rodriquez *et al.*, 2008), as well as uncapping of the physical association of single-strand binding proteins and telomerase from the extreme ends (Phatak *et al.*, 2006). It is notable that tumors over-expressing hPOT1 or TRF2 are resistant to G-quadruplex ligand (RHPS4) treatment (Salvati *et al.*, 2007). A plausible model for these events involves transient end-uncapping and exposure of the quadruplex–ligand complex at extreme telomere ends as the master signal that triggers the DNA damage response (Figure 6–1). Uncapping will also involves cross-talk between the components

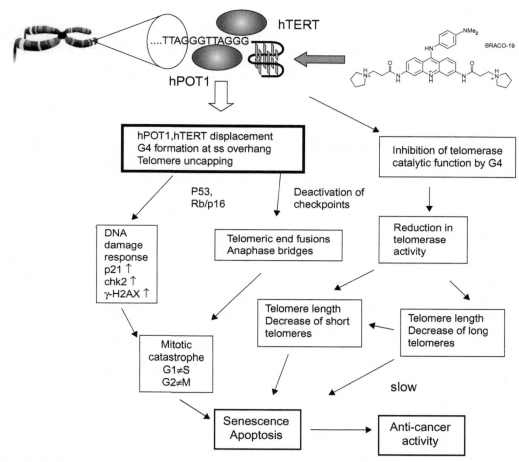

FIGURE 6–1 Schematic of the major short- and long-term cellular responses to G-quadruplex small molecules targeted to human cancer cell telomeres.

of the shelterin complex, with, for example, hPOT1 displacement dis-regulating telomerase function at the shelterin level since hPOT1 and Tpp1 facilitate telomere length regulation by telomerase (Denchi & de Lange, 2007). Q-FISH studies have shown that telomestatin is localized at telomeres during replication and importantly that normal cell telomere replication is unaffected in mouse embryonic fibroblast (i.e. untransformed) cell lines (Arnoult, Shin-Ya & Londoño-Vallejo, 2008). Analogous localization at telomere ends has been found for the quinolone ligand 360A in both normal and tumor cell (Granotier *et al.*, 2005).

The finding that cells exposed to RHPS4, which undergo telomere uncapping, then recruit and activate the PARP1 enzyme ((poly ADP ribose) polymerase-1) at telomeres (Salvati *et al.*, 2010), is another manifestation of the DNA damage response to quadruplex-binding ligands. It is of especial interest in view of the recent finding (Soldatenkov *et al.*, 2008) that PARP1 binds quadruplex DNA, and is enzymatically activated when it binds. It is plausible that the RHPS4–quadruplex complex is the recognition motif for PARP1 in these cells. The *in vivo* anti-tumor activity of RHPS4 is enhanced when used in conjunction with an established PARP1 inhibitor, although as with other combinations involving quadruplex ligands, the order of administration is critical.

G-quadruplex Ligands Show *in vivo* Anti-tumor Activity

The features common to most current quadruplex ligands, of several cationic charges and large hydrophobic surface area, do aid cellular uptake (probably by active transport mechanisms), but may also enable a high background of non-specific binding to cellular components, and are not consistent with oral bio-availability (though this may be an important future goal for the treatment of many cancers in the clinic). The three positive charges on the BRACO-19 molecule are probably a factor in the inability of this compound to penetrate larger tumors in both the UXF1138L and A431 xenograft models (Gowan *et al.*, 2002; Burger *et al.*, 2005) (Table 6–1). Compound AS1410 was devised (see Chapter 4) to have increased hydrophobicity compared to its parent compound BRACO-19 as a result of the modifications to the substituents at the 9-position. This resulted in an increase in plasma half-life from 1 to 2 h, although it did not result in a significant improvement in anti-cancer potency against tumor xenograft models.

BRACO-19 itself was found to elicit a rapid and progressive anti-cancer response in the UXF1138L xenograft (Figure 6–2), with tumor shrinkage of up to 96% and curative responses in some animals (Burger *et al.*, 2005). Immuno-staining of tumors with hTERT antibody demonstrated a major reduction in hTERT levels in treated tumors, consistent with the uncapping short-term mode of action model; the long-term telomere shortening outlined above suggests that both mechanisms are involved in the observed anti-cancer effects.

The *in vivo* xenograft data to date on telomeric quadruplex ligands (Table 6–1) are still limited in scope to a relatively small number of tumor models. However, they do suggest that telomeric quadruplex ligands may be useful for the treatment of solid tumors. Data on some of the more clinically challenging tumors such as renal and pancreatic would be welcome. There is very little data to date on hematological cancers apart from the encouraging results with telomestatin on the U937 leukemic lymphoma model (Tauchi *et al.*, 2006) of 80% tumor shrinkage. It may be that some of the present generation of quadruplex ligands have pharmacological properties that are favorable for leukemias and lymphomas, and further studies are warranted.

Notable findings with other tumors relevant to major unmet clinical need include that of single-agent activity for RHPS4 in a metastatic melanoma model, as well as in a melanoma line resistant to the platinum drug DDP (Leonetti *et al.*, 2008). RHPS4 appears able to penetrate significant tumor masses (Table 6–1), in accord with its single net positive charge combined with the relatively small size of this molecule and its favorable

Table 6–1   Selected *in vivo* data on quadruplex-binding ligands. Tumor responses have been estimated from survival curves and other available data. Reproduced with permission from Neidle (2010)

(a) Single-agent studies

G4 ligand	Xenograft model	Mean initial tumor size	Dosage in mg/kg	Tumor response	Days to complete response	Ref.
TMPyP4	MX-1 mammary tumor	100 mg*	10, 20; ip	Survival increase from 45 to 75%	60	Grand *et al.*, 2002
TMPyP4	PC-3 human prostate carcinoma	60 mg	40; ip	60% tumor shrinkage	18	Grand *et al.*, 2002
Telomestatin	U937 human lymphoma	1395 mm³	15	80% tumor shrinkage	21	Tauchi *et al.*, 2006
BRACO-19	UXF1138L human uterine carcinoma	68 mm³	2; ip	96% tumor shrinkage + some complete remissions	28	Burger *et al.*, 2005
BRACO-19	A431 human epithelial carcinoma	1080 mm³	2; ip	Not significant	–	Gowan *et al.*, 2002
Quarfloxin	MDA-MB-231 human breast cancer	>125 mm³	6.25, 15.5; iv	50% tumor shrinkage	37	Drygin *et al.*, 2009
Quarfloxin	MIA PaCa-2 human pancreatic cancer	>125 mm³	5; iv	59% tumor shrinkage	35	Drygin *et al.*, 2009
RHPS4	UXF1138L human uterine carcinoma	5 × 5 mm	5; oral	30% tumor shrinkage	28	Phatak *et al.*, 2007
RHPS4	M14, LP, LM melanoma	300–350 mg	10; ip	40–51% tumor weight reduction	15	Leonetti *et al.*, 2008
RHPS4**	CG5 breast carcinoma	300 mg	15; iv	75% tumor shrinkage	30	Salvati *et al.*, 2007

* Animals were initially treated with cyclophosphamide to minimize tumor burden.
** RHPS4 was reported in this study to have an anti-tumor effect in a number of other tumor types.

(b) Studies in combination with established anti-cancer drugs

G4 ligand	Dosage in mg/kg	Drug 2	Xenograft model	Initial tumor size	Tumor response	Ref.
BRACO-19	2	Paclitaxel	A431 epidermal carcinoma	1080 mm³	68% tumor shrinkage	Gowan *et al.*, 2002
AS1410	1	Cis-platinum	A549 lung carcinoma	10 mm³	ca 75% tumor shrinkage	Gunaratnam *et al.*, 2009a
RHPS4	5	Taxol	UXF1138L human uterine carcinoma	5 × 5 mm	Complete remissions	Phatak *et al.*, 2007
RHPS4*	10	Irinotecan	HCT116, HT29 colorectal carcinomas	300–350 mg	80% tumor weight reduction	Leonetti *et al.*, 2008

*A number of other combinations, with a range of anti-cancer drugs, were also reported in this study.

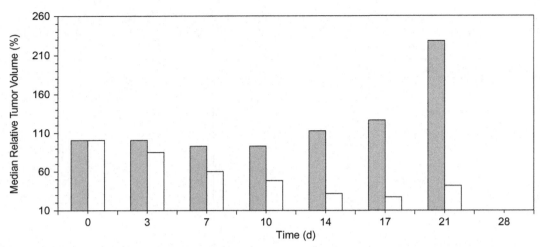

FIGURE 6–2 The response of the early-stage UXF1138LX vulval cell carcinoma xenograft model to BRACO-19 (adapted from Burger *et al.*, 2005). Solid bars represent growth of untreated tumors and shaded bars show responses of tumors treated i.p. with BRACO-19 at a dosage of 5 mg/kg over three weeks.

pharmacological features (Cookson, Heald & Stevens, 2005) that accord with the requirements of Lipinski's rules (Chapter 9) rather more than many of the current quadruplex ligands.

Data on two other quadruplex-binding ligands have also been included in this survey (Table 6–1). The porphyrin compound TMPyP4 (Chapter 4), which does bind with high affinity to a wide range of quadruplex nucleic acids, albeit with low selectivity, has been reported to show anti-cancer activity in MX-1 mammary tumors and PC-3 human prostate carcinomas (Grand *et al.*, 2002). Since TMPyP4 also binds strongly to duplex DNA, it is not clear if the anti-tumor activity is solely due to telomere targeting. Although quadruplexes in the promoter region of the *c-myc* oncogene have been suggested as a target for this compound, it is also an established telomerase inhibitor, so action at the telomere level should not be ruled out. *In vivo* data on the recently described quadruplex-binding fluoroquinolone derivative Quarfloxin (CX-3543) are included. It has been used in a number of recent clinical trials in human cancer so its broadly favorable pharmacological profile and low toxicity has relevance to other quadruplex ligands. This agent was initially suggested to be targeting a *c-myc* promoter quadruplex, but is now believed to function by selectively disrupting nucleolin/rDNA quadruplex complexes (Drygin *et al.*, 2009). It does not show the cellular behavior outlined above that is characteristic of a telomere targeting agent.

The observation of resistance to G-quadruplex ligands has not been reported in xenograft models, although several cell-based studies have investigated possible mechanisms for induced resistance. G-quadruplex ligands are active in ALT (Alternative Lengthening of Telomeres) cell lines, which do not express telomerase, as shown by the telomestatin-induced upcapping of telomeres in these lines (Temime-Smaali *et al.*, 2009), suggesting that reverse to an ALT phenotype is unlikely to be a resistance mechanism. In telomerase-positive cells, resistance to the triazine compound I2459 in A549 cells is accompanied by up-regulation of hTERT and increased telomerase expression (Gomez *et al.*, 2003), accompanied by resistance to telomere shortening. Over-expression of the apoptosis protein bcl-2 may also contribute (Douarre *et al.*, 2005), although it does not appear to be a major factor in resistance in the A549 lung carcinoma cell line.

It is encouraging for future clinical applications that several G-quadruplex ligands show synergistic activity *in vivo* (Table 6–1b) with conventional cytotoxic agents, such as cis-platinum, taxol and camptothecin derivatives (Gowan *et al.*, 2002; Phatak *et al.*, 2006; Leonetti *et al.*, 2008; Gunaratnam *et al.*, 2009a), although the detailed mechanism of this effect remains to be established. The order in which the drugs are administered appears to be an important determinant of whether a particular combination is synergistic or antagonistic. It is also entirely possible that quadruplex-binding ligands can have multiple quadruplex targets (telomeric and genomic), which could confer therapeutic advantage.

Dual targeting has been reported for a substituted naphthalene diimide, which interacts with quadruplexes in the promoter region of the *c-kit* oncogene that is disregulated in gastrointestinal cancer cells and down-regulates *c-kit* expression while also affecting telomere maintenance. Inhibition of *c-kit* expression and telomerase activity take place at the ligand concentrations required to halt cell growth and proliferation (Gunaratnam *et al.*, 2009b). Dual targeting of oncogenic genomic and telomeric quadruplexes by G-quadruplex ligands may be an attractive strategy for the future. It may also be what is taking place with existing agents that have limited selectivity for a particular quadruplex, which may even confer therapeutic advantage.

References

Arnoult, N., Shin-Ya, K., Londoño-Vallejo, J.A., 2008. Studying telomere replication by Q-CO-FISH: the effect of telomestatin, a potent G-quadruplex ligand. Cytogenet. Genome Res. 122, 229–236.

Binz, N., Shalaby, T., Rivera, P., Shin-ya, K., Grotzer, M.A., 2005. Telomerase inhibition, telomere shortening, cell growth suppression and induction of apoptosis by telomestatin in childhood neuroblastoma cells. Eur. J. Cancer 41, 2873–2881.

Burger, A.M., Dai, F., Schultes, C.M., Reszka, A.P., Moore, M.J., Double, J.A., et al., 2005. The G-quadruplex-interactive molecule BRACO-19 inhibits tumor growth, consistent with telomere targeting and interference with telomerase function. Cancer Res. 65, 1489–1496.

Casagrande, V., Salvati, E., Alvino, A., Bianco, A., Ciammaichella, A., D'Angelo, C., et al., 2011. N-cyclic bay-substituted perylene G-quadruplex ligands have selective antiproliferative effects on cancer cells and induce telomere damage. J. Med. Chem. 54, 1140–1156.

Cookson, J.C., Dai, F., Smith, V., Heald, R.A., Laughton, C.A., Stevens, M.F.G., et al., 2005. Pharmacodynamics of the G-quadruplex-stabilizing telomerase inhibitor 3, 11-difluoro-6,8, 13-trimethyl-8H-quino[4,3,2-kl]acridinium methosulfate (RHPS4) in vitro: activity in human tumor cells correlates with telomere length and can be enhanced, or antagonized, with cytotoxic agents. Mol. Pharmacol. 68, 1551–1558.

Cookson, J.C., Heald, R.A., Stevens, M.F., 2005. Antitumor polycyclic acridines. 17. Synthesis and pharmaceutical profiles of pentacyclic acridinium salts designed to destabilize telomeric integrity. J. Med. Chem. 48, 7198–7207.

De Cian, A., Lacroix, L., Douarre, C., Temime-Smaali, N., Trentesaux, C., Riou, J.-F., et al., 2008. Targeting telomeres and telomerase. Biochimie 90, 131–155.

Denchi, E.L., de Lange, T., 2007. Protection of telomeres through independent control of ATM and ATR by TRF2 and POT1. Nature 448, 1068–1071.

Douarre, C., Gomez, D., Morjani, H., Zahm, J.M., O'Donohue, M.F., Eddabra, L., et al., 2005. Overexpression of Bcl-2 is associated with apoptotic resistance to the G-quadruplex ligand 12459 but is not sufficient to confer resistance to long-term senescence. Nucleic Acids Res. 33, 2192–2203.

Drygin, D., Siddiqui-Jain, A., O'Brien, S., Schwaebe, M., Lin, A., Bliesath, J., et al., 2009. Anticancer activity of CX-3543: a direct inhibitor of rRNA biogenesis. Cancer Res. 71, 1418–1430.

Gomez, D., Aouali, N., Renaud, A., Douarre, C., Shin-Ya, K., Tazi, J., et al., 2003. Resistance to senescence induction and telomere shortening by a G-quadruplex ligand inhibitor of telomerase. Cancer Res. 63, 6149–6153.

Gomez, D., O'Donohue, M.F., Wenner, T., Douarre, C., Macadré, J., Koebel, P., et al., 2006a. The G-quadruplex ligand telomestatin inhibits POT1 binding to telomeric sequences in vitro and induces GFP-POT1 dissociation from telomeres in human cells. Cancer Res. 66, 7908–7912.

Gomez, D., Wenner, T., Brassart, B., Douarre, C., O'Donohue, M.F., El Khoury, V., et al., 2006b. Telomestatin-induced telomere uncapping is modulated by POT1 through G-overhang extension in HT1080 human tumor cells. J. Biol. Chem. 281, 38721–38729.

Gowan, S.M., Harrison, J.R., Patterson, L., Valenti, M., Read, M.A., Neidle, S., et al., 2002. A G-quadruplex-interactive potent small-molecule inhibitor of telomerase exhibiting in vitro and in vivo antitumor activity. Mol. Pharmacol. 61, 1154–1162.

Grand, C.L., Han, H., Munoz, R.M., Weitman, S., Von Hoff, D.D., Hurley, L.H., et al., 2002. The cationic porphyrin TMPyP4 down-regulates c-myc and human telomerase reverse transcriptase expression and inhibits tumor growth in vivo. Mol. Cancer Ther. 1, 565–573.

Granotier, C., Pennarun, G., Riou, L., Hoffschir, F., Gauthier, L.R., De Cian, A., et al., 2005. Preferential binding of a G-quadruplex ligand to human chromosome ends. Nucleic Acids Res. 33, 4182–4190.

Gunaratnam, M., Greciano, O., Martins, C., Reszka, A.P., Schultes, C.M., Morjani, H., et al., 2007. Mechanism of acridine-based telomerase inhibition and telomere shortening. Biochem. Pharmacol. 74, 679–689.

Gunaratnam, M., Green, C., Moreira, J.B., Moorhouse, A.D., Kelland, L.R., Moses, J.E., et al., 2009a. G-quadruplex compounds and cis-platin act synergistically to inhibit cancer cell growth in vitro and in vivo. Biochem. Pharmacol 78, 115–122.

Gunaratnam, M., Swank, S., Haider, S.M., Galesa, K., Reszka, A.P., Beltran, M., et al., 2009b. Targeting human gastrointestinal stromal tumor cells with a quadruplex-binding small molecule. J. Med. Chem. 52, 3774–3783.

Hemann, M.T., Strong, M.A., Hao, L.Y., Greider, C.W., 2001. The shortest telomere, not average telomere length, is critical for cell viability and chromosome stability. Cell 107, 67–77.

Incles, C.M., Schultes, C.M., Kempski, H., Koehler, H., Kelland, L.R., Neidle, S., 2004. A G-quadruplex telomere targeting agent produces p16-associated senescence and chromosomal fusions in human prostate cancer cells. Mol. Cancer Ther. 3, 1201–1206.

Leonetti, C., Scarsella, M., Riggio, G., Rizzo, A., Salvati, E., D'Incalci, M., et al., 2008. G-Quadruplex ligand RHPS4 potentiates the antitumor activity of camptothecins in preclinical models of solid tumors. Clin. Cancer Res. 14, 7284–7291.

Neidle, S., 2010. The current status of telomeric G-quadruplexes as therapeutic targets in human cancer. FEBS J. 277, 1118–1125.

Pennarun, G., Granotier, C., Hoffschir, F., Mandine, E., Biard, D., Gauthier, L.R., et al., 2008. Role of ATM in the telomere response to the G-quadruplex ligand 360A. Nucleic Acids Res. 36, 1741–1754.

Phatak, P., Burger, A.M., 2007. Telomerase and its potential for therapeutic intervention. Brit. J. Pharmacol. 152, 1003–1011.

Phatak, P., Cookson, J.C., Dai, F., Smith, V., Gartenhaus, R.B., Stevens, M.F.G., et al., 2007. Telomere uncapping by G-quadruplex ligand RHPS4 inhibits clonogenic tumour cell growth *in vitro* and *in vivo* consistent with a cancer stem cell targeting mechanism. Brit. J. Cancer 96, 1223–1233.

Riou, J.-F., 2004. G-quadruplex interacting ligands targeting the telomeric G-overhang are more than simple telomerase inhibitors. Curr. Med. Chem. 4, 439–443.

Rizzo, A., Salvati, E., Porru, M., D'Angelo, C., Stevens, M.F., D'Incalci, M., et al., 2009. Stabilization of quadruplex DNA perturbs telomere replication leading to the activation of an ATR-dependent ATM signaling pathway. Nucleic Acids Res. 37, 5353–5364.

Rodriguez, R., Müller, S., Yeoman, J.A., Trentesaux, C., Riou, J.-F., Balasubramanian, S., 2008. A novel small molecule that alters shelterin integrity and triggers a DNA-damage response at telomeres. J. Amer. Chem. Soc. 130, 15758–15759.

Salvati, E., Leonetti, C., Rizzo, A., Scarsella, M., Mottolese, M., Galati, R., et al., 2007. Telomere damage induced by the G-quadruplex ligand RHPS4 has an antitumor effect. J. Clin. Investig. 117, 3236–3247.

Salvati, E., Scarsella, M., Porru, M., Rizzo, A., Iachettini, S., Tentori, L., et al., 2010. PARP1 is activated at telomeres upon G4 stabilization: possible target for telomere-based therapy. Oncogene 29, 6280–6293.

Shammas, M.A., Koley, H., Beer, D.G., Li, C., Goyal, R.K., Munshi, N.C., 2004. Growth arrest, apoptosis, and telomere shortening of Barrett's-associated adenocarcinoma cells by a telomerase inhibitor. Gastroenterology 126, 1337–1346.

Soldatenkov, V.A., Vetcher, A.A., Duka, T., Ladame, S., 2008. First evidence of a functional interaction between DNA quadruplexes and poly(ADP-ribose) polymerase-1. ACS Chem. Biol. 3, 214–219.

Tahara, H., Shin-Ya, K., Seimiya, H., Yamada, H., Tsuruo, T., Ide, T., 2006. G-Quadruplex stabilization by telomestatin induces TRF2 protein dissociation from telomeres and anaphase bridge formation accompanied by loss of the 3′ telomeric overhang in cancer cells. Oncogene 25, 1955–1966.

Tauchi, T., Shin-ya, K., Sashido, G., Sumi, M., Nakajima, A., Shimamoto, T., et al., 2003. Activity of a novel G-quadruplex-interactive telomerase inhibitor, telomestatin (SOT-095), against human leukemia cells: involvement of ATM-dependent DNA damage response pathways. Oncogene 22, 5338–5347.

Tauchi, T., Shin-ya, K., Sashido, G., Sumi, M., Okabe, S., Ohyashiki, J.H., et al., 2006. Telomerase inhibition with a novel G-quadruplex-interactive agent, telomestatin: in vitro and in vivo studies in acute leukemia. Oncogene 25, 5719–5725.

Temime-Smaali, N., Guittat, L., Sidibe, A., Shin-ya, K., Trentesaux, C., Riou, J.-F., 2009. The G-quadruplex ligand telomestatin impairs binding of topoisomerase IIIα to G-quadruplex-forming oligonucleotides and uncaps telomeres in ALT cells. PLoS One 4, 6919.

Zhou, W.-J., Deng, R., Zhang, X.-Y., Feng, G.-K., Gu, L.-Q., Zhu, X.-F., 2009. G-quadruplex ligand SYUIQ-5 induces autophagy by telomere damage and TRF2 delocalization in cancer cells. Mol. Cancer Ther. 8, 3203–3213.

7

Genomic Quadruplexes as Therapeutic Targets

Introduction

The realization is not new that regions of the genome beyond the telomere have potential quadruplex-forming ability, and that quadruplex-binding proteins exist (see Simonsson, 2001 for an early review of this field). Suitable G-tract sequences were identified in a number of genes (such as the retinoblastoma gene: Murchie & Lilley, 1992; Xu & Sugiyama, 2006), and in breakpoint regions well before the human genome was fully sequenced, as outlined in Chapter 1. Sequences capable of forming intramolecular quadruplexes within the promoter region of the *c-myc* oncogene were the first to be systematically studied (Simonsson, Pecinka & Kubista, 1998), and are still of much interest, for several reasons:

1. *c-myc* is a ubiquitous control oncogene, playing a key role in many human cancers.
2. The MYC gene product is unstable and only transiently structured, so it is a virtually undruggable target and alternative approaches to inhibiting its function are highly desirable.
3. Targeting the *c-myc* quadruplexes with small-molecule compounds has become a paradigm for the rapidly expanding field of promoter quadruplex targeting (Brooks & Hurley, 2010) which can be systematically approached now that full sequence data are available for human and many other genomes. An increasing number of such quadruplexes have now been characterized and the effects of small molecules on them are being evaluated.

The concept is that transcription at the individual gene level can be down-regulated by the action of a small molecule that stabilizes the quadruplex; the resulting quadruplex–ligand complex would thus present an impediment to RNA polymerase reading and to binding of transcription factors such as SP1. This necessitates that the two strands of the duplex promoter are separated (Figure 7–1), with the G-rich strand forming a quadruplex structure, and the C-rich strand being either (relatively) unstructured, or forming a mobile i-motif arrangement (Simonsson, Pribylova & Vorlickova, 1999) with the quadruplex then stabilized by a ligand (Rangan, Fedoroff & Hurley, 2001; Brooks, Kendrick & Hurley, 2010). The i-motif has been previously well established as forming in C-rich DNA sequences, albeit with a low pH requirement in order for cytosine bases to be

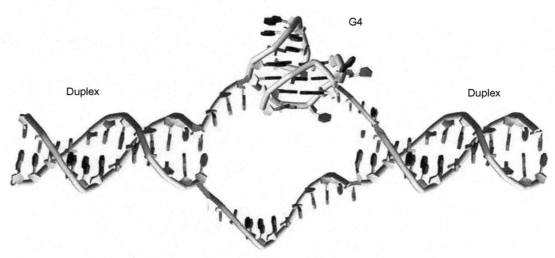

FIGURE 7–1 A schematic view of the separation of two strands of duplex DNA to form a quadruplex structure on one, the G-rich strand. The second C-rich strand is shown in a single-strand form. The quadruplex shown is that of *c-kit1* (see Section 7.3).

protonated and the characteristic C^+–C base pairing to be formed (see, for example, Phan, Guéron & Leroy, 2000; Phan & Mergny, 2002; Dai *et al.*, 2009). The energetics of strand separation and the topological constraints of the double helix (Brooks & Hurley, 2009) both need to be overcome prior to quadruplex formation, although separation does occur during replication when higher-order structures may be transiently formed. There is some biophysical evidence that the equilibrium can favor the formation of quadruplex structure within an environment of an appropriate G-rich long duplex sequence, for example from single-molecule studies of 100 base pair DNA sequences containing *c-myc* or *c-kit* quadruplexes embedded within their natural genomic environments (Shirude *et al.*, 2007; Shirude, Ling & Balasubramanian, 2008). The consequences of small molecule binding to and further stabilizing the quadruplex arrangement are analogous to telomere targeting in that in both instances the effect on the relevant polymerases (RNA polymerase for transcription and telomerase for telomeres) is indirect. Attention has wholly focused to date on intramolecular quadruplexes, but especially when a promoter site contains > four G-tracts, then the formation of bimolecular quadruplexes is at least theoretically possible.

The issue of quadruplex sequence prevalence in promoter regions of the human genome has been systematically studied by bioinformatics methods (Huppert & Balasubramanian, 2007; Du, Zhao & Li, 2009), following on from earlier studies (summarized in Chapter 1), that have examined all regions, coding and non-coding, in the total human genome sequence. This more focused question requires as prerequisite a definition of the established promoter sequences—available, for example, through the annotated ENSEMBL genomic browser (www.ensembl.org). The searches were conducted

(Huppert & Balasubramanian 2007) with the same generalized quadruplex sequence as has been used in the initial genomic searches, and were restricted to regions within 1 kilobase upstream of the 5′ end of transcription start sites. A total of 14,769 putative quadruplex sequences were located in the 19,268 human genes known at that time, and over 40% of these genes contain >1 such sequence. The probability of finding such a sequence is correlated with proximity to the transcription start site, as well as its location within a nuclease hypersensitive site. A significant factor in this enrichment is that the G-rich binding site for the SP1 general transcription factor also occurs in this region close to the transcription start site, so unsurprisingly many of these putative quadruplex sequences also contain an SP1 site (Todd & Neidle, 2008). Many of the genes containing promoter quadruplexes are involved in proliferation, and thus include the majority of human oncogenes. Over-representation of putative quadruplex sites has also been found in the first intron of ~16% of human genes, and is strongly conserved in a number of species (Eddy & Maizels, 2008). An additional layer of complexity has been added by the finding (Eddy *et al.*, 2011) that transcription of human genes can pause immediately downstream of the transcription start site, and that this correlates with the occurrence of putative quadruplex sequences within the first intron. Experimental data relevant to the concept that promoter quadruplexes can be involved in transcriptional regulation have been obtained using a human genome-wide microarray approach (Verma *et al.*, 2008, 2009), following exposure of HeLa and A549 cancer cells to the porphyrin compound TMPyP4. A note of caution is needed in interpreting such studies since although the transcription of such genes as *c-myc* were clearly down-regulated, this ligand is non-selective and binds equally well to quadruplex and duplex DNA. Another complicating factor when studying promoter quadruplex sequences is that many contain more than four G-tracts, suggesting the possibility of several inter-converting quadruplex structures. This has been observed for *c-myc* and several other targets (Brooks, Kendrick & Hurley, 2010). Finally it is worth noting that promoter quadruplex sequences are not confined to the human genome, and could find utility as targets in anti-infective chemotherapy. They are widely distributed in the yeast *Saccharomyces cerevisiae* (Hershman *et al.*, 2008; Capra *et al.*, 2010), as well as in many prokaryotic organisms including *E. coli* (Rawal *et al.*, 2006).

Targeting promoter quadruplexes with small molecules can be aiming for selectivity at the individual gene level, in which case the ligand must be able to discriminate between different quadruplexes (Balasubramanian, Hurley & Neidle, 2011). The effect would be somewhat amplified if the target gene is strongly expressing and is functionally important in a particular cell type, but is still a major challenge. As will be discussed below, structural information is available to date on only a very small number of promoter quadruplexes (Table 7–1). The data do suggest that some of them will have sufficient structural distinctiveness to make them druggable targets, with the *c-kit1* and *c-kit2* quadruplexes being good examples. It is not feasible to experimentally scan all potential promoter quadruplexes with current methodology, or even a significant subset. So a realistic strategy is to preselect a target on the basis of prior target validation information and

Table 7–1 Structures of genomic quadruplexes available from the Protein Data Bank. All have been determined by NMR methods except for entry 3QXR, which is a crystal structure

Gene	Sequence	PDB id	Reference
hTERT	$AG_3IAG_3CTG_3AG_3C$	2KZD	Lim *et al.*, 2010
hTERT	$AIG_3AG_3ICTG_3AG_3C$	2KZE	Lim *et al.*, 2010
c-kit2	$CG_3CG_3CGCGAG_3AG_3T$	2KYO	Kuryavyi, Phan & Patel, 2010
c-kit2	$CG_3CG_3CGCTAG_3AG_3T$	2KYP	Kuryavyi, Phan & Patel, 2010
c-kit2	$CG_3CG_3CACGAG_3AG_3T$	2KQG	Hsu *et al.*, 2009
c-kit2	$CG_3CG_3CGCGAG_3AG_3T$	2KQH	Hsu *et al.*, 2009
c-kit1	$AG_3AG_3CGCTG_3AG_2AG_3$	2O3M	Phan *et al.*, 2007
c-kit1 (xtl)	$AG_3AG_3CGCTG_3AG_2AG_3$	3QXR	Wei, Parkinson & Neidle, 2011
bcl-2	$G_3CGCG_3AG_2A_2T_2G_3CG_3$	2F8U	Dai *et al.*, 2006
c-myc	$TGAG_3TG_3TAG_3TG_3TAA$	1XAV	Ambrus *et al.*, 2005
c-myc+TMPyP4	$TGAG_3TG_2IGAG_3TG_4A_2G_2$	2A5R	Phan *et al.*, 2005
c-myc	$TGAG_3TG_2IGAG_3TG_4A_2G_2$	2A5P	Phan *et al.*, 2005

then progress sequentially through a series of queries each requiring clarification before proceeding through to a drug discovery program:

- Does a plausible quadruplex sequence exist within the promoter?
- If so, is it stable under physiological conditions? Biophysical studies will answer this question.
- Is the quadruplex sequence unique? Bioinformatics methodology is available for this (see Chapter 1).

If the quadruplex looks to be a plausible target at this point then a full structure determination by NMR or crystallography is warranted at this stage, to determine whether the structure has unusual or unique fold/loops and other structural features.

It is also not possible to ascertain the global selectivity of any ligands reported to date that have the intent of targeting individual promoter quadruplexes, although the structural simplicity of almost if not all these ligands suggests that they are highly unlikely to be specific for a single target. Indeed the likely lack of selectivity may even confer therapeutic advantage given the multiple dis-regulation and interdependence of signaling pathways in the majority of advanced cancers.

Quadruplexes in the *c-myc* Promoter

The *c-myc* gene is a master controller oncogene, over-expressed in over 80% of human cancers (Pelengaris, Khan & Evan, 2002; Dang, 2010). The encoded MYC protein is a transcription factor, which when mutated or up-regulated contributes to dis-regulated growth control and hence to oncogenesis. The NHE III$_1$ element in the *c-myc* promoter has nuclease hypersensitivity, and controls ~90% of the transcription of this gene. The 27 nucleotide sequence in this element contains six G-tracts, of which five contain three or more Gs:

5'-TGGGGAGGGTGGGGAGGGTGGGGAAGG

The occurrence of >four G-tracts in a promoter sequence is common and almost the norm, and indeed this sequence is embedded in a longer G-rich region of the *c-myc* promoter with other G-tracts. This feature of multiple G-tracts implies structural ambiguity in defining an individual quadruplex, as well as indicating the high likelihood of dynamic behavior, switching between quadruplexes and between folds along the length of the region.

Quadruplex formation in the *c-myc* sequence has been unequivocally demonstrated by a variety of footprinting and biophysical methods (Siddiqui-Jain *et al.*, 2002), and the role of the sequence in the function of this gene determined by an extensive series of mechanistic studies (Dexheimer *et al.*, 2009; Sun & Hurley, 2009; Brooks & Hurley, 2010; González & Hurley, 2010a, b). An early finding was that formation of quadruplex(es) in the NHE III$_1$ element inhibits transcription, as predicted. Furthermore, proteins involved in both promoting (the highly abundant nuclear protein nucleolin) and resolving quadruplex formation (the NM23H2 protein) have been characterized, providing strong circumstantial evidence for the natural involvement of quadruplex folding in the regulation of *c-myc* expression. Thus nucleolin represses *c-myc* transcription while NM23H2 promotes it.

There are structural ambiguities present in the 27-mer NHE III$_1$ element as a result of the presence of six G-tracts. *A priori* it is not apparent which intramolecular quadruplexes are formed and which are more physiologically relevant for *c-myc* transcription. This question has been addressed by systematic mutagenesis and chemical footprinting studies which indicated that the first four 5' G-tracts are the most significant ones (Brooks, Kendrick & Hurley, 2010). NMR structure determinations (Table 7–1) have been performed on several regions of the complete 27-mer, which itself unsurprisingly, at least in dilute solution, forms multiple quadruplex species (Phan, Modi & Patel, 2004; Hatzakis, Okamoto & Yang, 2010). Two quadruplexes comprising G-tracts 1,2,4,5 and 2,3,4,5 have been analyzed by NMR structure determination methods (Phan, Modi & Patel, 2004). The former has the two halves (G-tracts 1,2 and 4,5) linked by an artificial T$_4$ sequence. Both sequences form very stable parallel-stranded quadruplexes, with the 2,3,4,5 one being especially resilient and having a T$_m$ >80°. A second analysis has been reported (Ambrus *et al.*, 2005) of the sequence comprising G-tracts 2,3,4,5, but with two G→T mutations to enhance the stability of a single conformer. This sequence forms an equivalent parallel quadruplex (Figure 7–2) with both containing two single-nucleotide (T) and one two-nucleotide (GA) propeller loops. The G-tract 1,2,4,5 parallel quadruplex

FIGURE 7–2 One of the ensemble of NMR structures determined for the parallel-form *c-myc* promoter quadruplex (Ambrus *et al.*, 2005), shown in cartoon form.

has single-nucleotide propeller loops (A and T), as well as a long T_5A one. This thus provides important information relevant to other promoter quadruplexes, namely that the presence of two single-nucleotide loops (a common occurrence) confers a strong tendency for a parallel topology, even if the third loop is long. A 24-nucleotide variant of the G-tract 2,3,4,5 quadruplex has been studied by NMR (Phan *et al.*, 2005) and also contains the 3′-terminal GG sequence, i.e. is essentially five G-tracts in length. This GG sequence plays a significant structural role and is folded back so that one of these two guanines is inserted into the terminal G-quartet. Overall the structure retains the parallel topology observed for all *c-myc* quadruplexes but in terms of the platforms created by flanking sequences it is distinctive from the other 2,3,4,5 quadruplexes.

The *c-myc* quadruplex system has been a well-studied target for small-molecule ligands (Peng *et al.*, 2010), although few are specific for these quadruplexes. The porphyrin compound TMPyP4 binds exceptionally tightly to the quadruplexes and reduces *c-myc* expression both in a cell-free system and in cells (Grand *et al.*, 2002; Hurley *et al.*, 2006), whereas its stereochemically restricted isomer TMPyP2 binds much less tightly and does not show these biological effects. A *c-myc* quadruplex–TMPyP4 complex has been analyzed by NMR methods (Table 7–1: Phan *et al.*, 2005) and shows the ligand bound onto a terminal G-quartet face, as expected (Figure 7–3a, b). The ligand is kept in place by a bridge comprising the first two 5′ bases (T and G) and A12. TMPyP4 binds equally well to many other quadruplexes such as human telomeric ones and also binds almost as effectively to duplex DNA, and indeed to other nucleic acid structures (Ren & Chaires, 1999). Other ligands that have been shown to bind to *c-myc* quadruplexes are shown in Figure 7–4. These show high structural diversity and there are no obvious structure–activity relationships between them, although several broad features are apparent, which are consistent with the binding site in the TMPyP4 complex (Figure 7–5). The G-quartet surface in this structure appears to be accessible to a range of ligand shapes

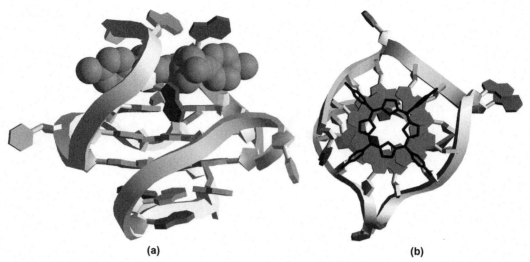

(a)   (b)

FIGURE 7–3 Two views of the NMR structure of the *c-myc* quadruplex complex with the porphyrin compound TMPyP4 (Phan *et al.*, 2005), **(a)** showing the TMPyP4 molecule in space-filling representation. **(b)** A view down onto the G-quartets with TMPyP4 in stick representation, showing the positioning of the pyridyl groups in the quadruplex grooves.

and sizes, both to ligands with large diameters and to those ligands comprising extended polycyclic aromatic chromophores. A few of these compounds are highlighted here:

1. Cyclic furanoid oligopeptides (Figure 7–4a) are overall similar in size to TMPyP4 (Figure 7–6), although unlike TMPyP4 they do not bind significantly to duplex DNA. They down-regulate *c-myc* transcription by 90% (Agarwal *et al.*, 2010) in a reporter assay. Apoptosis was observed on treatment of HeLa cells, which is consistent with *c-myc* targeting. These compounds can be considered to be analogous in structural terms to the potent quadruplex-binding natural product telomestatin (Chapter 5). The telomestatin molecule is slightly larger in surface area than the furanoid oligopeptides, and molecular modeling suggests that it can overlap optimally with a G-quartet in the *c-myc* structure (Figure 7–7). A closely related bis-amide derivative of telomestatin, termed S2T1-6OTD, has provided a further justification for targeting the *c-myc* quadruplexes. This compound binds preferentially (Shalaby *et al.*, 2010) to a 22-mer from the *c-myc* promoter region, contrasting with its weak affinity for a human telomeric sequence and no detectable affinity for duplex DNA. Inhibition of transcription was observed by significant reductions in *c-myc* mRNA and protein expression in a panel of childhood medulloblastoma cells dosed with the compound, which also showed long-term growth arrest at subtoxic doses. Further evidence of the involvement of the NHE III$_1$ sequence was provided by experiments using closely related cell lines with translocation breakpoints within this sequence (so quadruplex

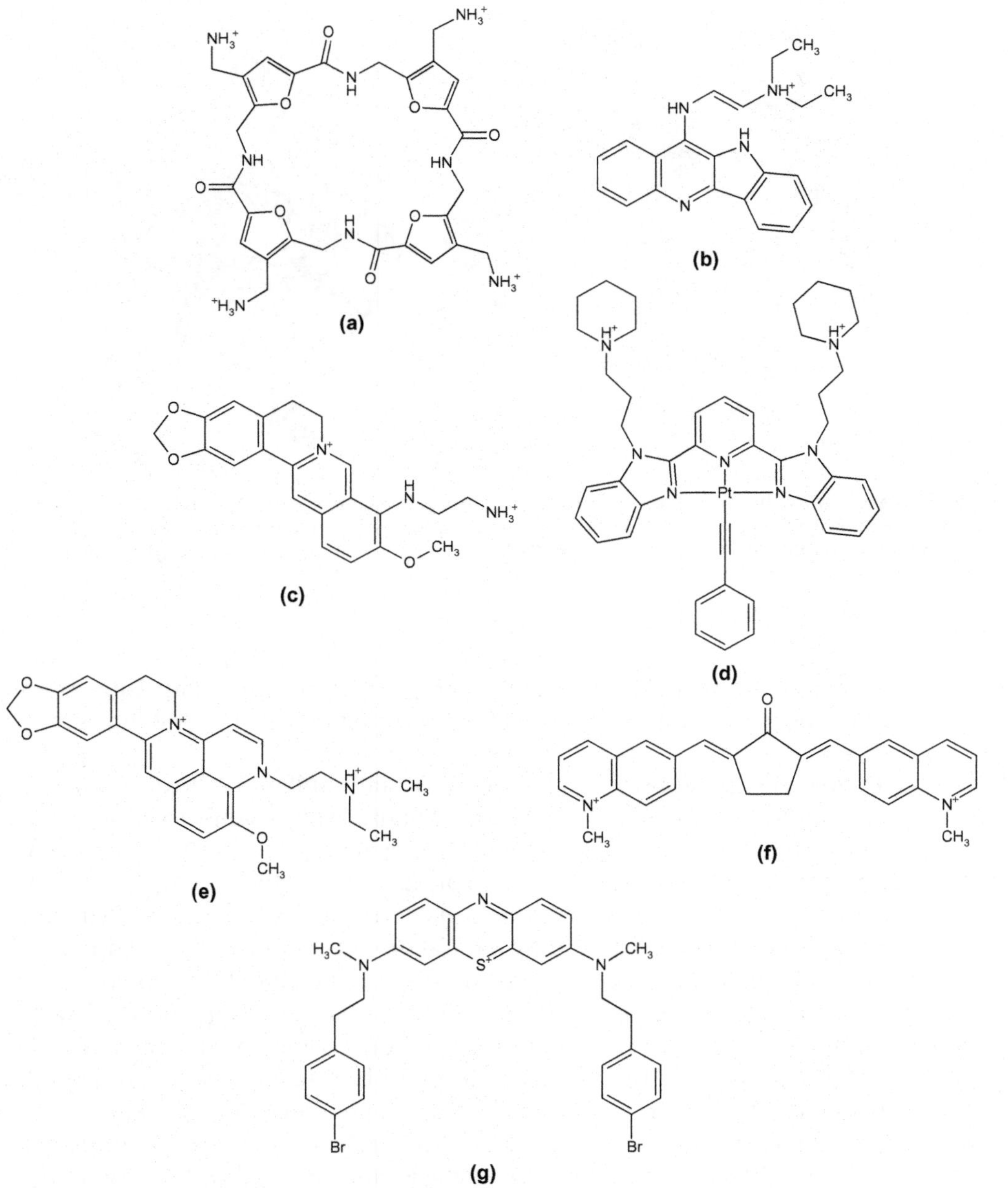

FIGURE 7–4 (a–g) Small-molecule ligands that have been reported to bind to *c-myc* quadruplexes. See the text for details.

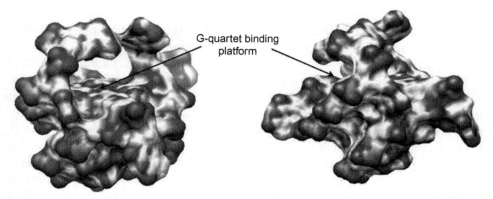

FIGURE 7–5 Two views of the solvent-accessible surface of the *c-myc* quadruplex TMPyP4 complex, with the ligand removed. The bridging nucleotides that cap the ligand binding site are seen, together with the G-quartet platform.

FIGURE 7–6 Superposition of the TMPyP4 structure (shown in dark shading) on that of a cyclic furanoid oligopeptide (Agarwal *et al.*, 2010).

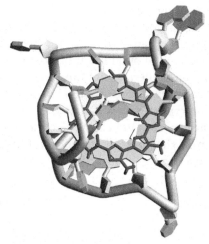

FIGURE 7–7 A molecular model of telomestatin bound to the *c-myc* quadruplex, using the NMR structure of the TMPyP4 complex (Phan *et al.*, 2005) as a starting point. Note the excellent overlap between the ligand (in stick representation) and the G-quartet.

formation is no longer possible), when no significant effects on *c-myc* transcription were found.

2. Bis-bromophenylethyl derivatives of the phenothiazine compound methylene blue (Figure 7–4b) were found by a virtual screening approach using the FDA (USA Food and Drug Administration) database of approved drugs together with a model for

the *c-myc* quadruplex structure (Chan *et al.*, 2011). This study thus took an existing approved drug as a starting point, probably the first time that this approach has been taken for quadruplex-binding ligands. Interestingly, this study did not use one of the existing *c-myc* NMR structures, but instead built a homology model from the human telomeric DNA parallel quadruplex crystal structure (Parkinson, Lee & Neidle, 2002), thereby avoiding any potential pitfalls associated with choosing from among the ensemble of NMR structures. This model has been used previously in studies of several other ligands (Figure 7–4a) such as quinolone derivatives (Ma *et al.*, 2011) and a platinum Schiff base complex (Wang *et al.*, 2010). Methylene blue was found to be a plausible hit compound and structure-based lead optimization led to the bis-bromophenyl derivatives (Figure 7–4b). The addition of the two side-chains led to markedly enhanced quadruplex affinity and selectivity. The lead compound, which inhibits proliferation in a cancer cell line, also inhibits *c-myc* transcription in a plasmid reporter assay transfected into cells.

The study on methylene blue derivatives (Chan *et al.*, 2011) indicates that structure-based design can play a role in finding novel new ligand leads (see also Chapter 9); whether a homology model or one of the NMR structures (Figure 7–8) is a better starting point remains to be established. The unequivocal demonstration that activity in cells, and *in vivo* of such compounds is a consequence of targeting the *c-myc* quadruplex, is a crucial challenge that has to be overcome for all promoter quadruplex ligands. Recent studies suggest that there is an increasing awareness of these issues and the advent of robust reporter assays that can be used in target cells is providing a firm validation of the approach. One can thus conclude that there is real progress in the first crucial stage towards the goal of a promoter quadruplex ligand small molecule being trialed for human use.

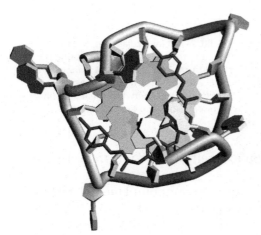

FIGURE 7–8 A molecular model of one of the bis-bromophenylethyl derivatives of the phenothiazine compound methylene blue (Chan *et al.*, 2011) bound to the *c-myc* quadruplex, also using the NMR structure of the TMPyP4 complex (Phan *et al.*, 2005) as a starting point. The bromophenylethyl side-chains may also adopt alternative conformations in which they are situated within the grooves.

Quadruplexes in the *c-kit* Promoter

The *c-kit* proto-oncogene, by contrast with *c-myc*, is of more restricted proven clinical relevance. Although the KIT protein, which is a tyrosine kinase receptor, is involved in a number of growth regulatory pathways, its dis-regulation has been unequivocally implicated in only a few cancers (Fletcher & Rubin, 2007), the most well studied being gastrointestinal stromal tumors (GISTs). Constitutive activation of *c-kit* expression is probably the primary causative event in this traditionally hard-to-treat cancer. By contrast with MYC, the KIT protein can be successfully targeted by a number of selective small-molecule compounds (notably by Gleevec, which was rapidly approved for human clinical use in GIST patients following on from extraordinary responses in clinical trials) although in common with other kinase-inhibitor approaches, active-site mutations inevitably occur, leading to clinical resistance (Tuveson *et al.*, 2001). So there continues to be a clinical need for an alternative non-kinase targeting approach that circumvents this major resistance problem.

The core *c-kit* promoter region contains two sequences of G-tracts separated by 31 nucleotides. The *c-kit* quadruplex sequences occur between –87 and –109 base pairs (*c-kit1*: Rankin *et al.*, 2005) and between –140 and –160 base pairs (*c-kit2*: Fernando *et al.*, 2006) upstream of the transcription start site. Both sequences have unique occurrence in the human genome (Todd *et al.*, 2007). Biophysical methods have shown that both form stable quadruplex structures. Unusually each of these includes just four G-tracts rather than the multiple G-tracts commonly found in promoter regions. A number of mutants in the non G-tract regions of the *c-kit1* quadruplex have been examined (Figure 7–9). Biophysical studies on these mutant sequences showed that surprisingly they did not form stable quadruplexes in solution, indicating that the *c-kit1* quadruplex has highly specific requirements for stability. These findings also lead to the more general conclusion that possession of the requisite number of G-tracts is not by itself sufficient for a stable quadruplex to be formed, suggesting that the total number of stable quadruplexes encoded in the human genome is in reality much less than the total number of quadruplex sequences (Chapter 1).

The molecular structures of both *c-kit* quadruplexes have been determined by NMR and in the case of *c-kit1*, by crystallography as well. The native 22-mer *c-kit1* sequence forms a single species in solution (Phan *et al.*, 2007), which has a unique topology (Figure 7–10) that is preserved in the crystal structure (Wei *et al.*, to be published). The core of the quadruplex comprises three stacked G-quartets, with one of the "loop" guanines flipping back to be part of one quartet. The net effect is to produce a parallel fold with

(a)	c-kit native	AGGGAGGGCGCTGGGAGGAGGG
(b)	mod 1	AGGGAGGGCGCTGGGCGCTGGG
(c)	mod 2	AGGGAGGGCGCTGGGAGGAGGG
(d)	mod 3	AGGGAGGGCGCTGGGCGGCGGG

1   5   10   15   20

FIGURE 7–9 The primary sequence of the *c-kit1* quadruplex, together with three mutated sequences that do not form stable quadruplexes (Rankin *et al.*, 2005). The G-tract sequences are shown with lighter lettering.

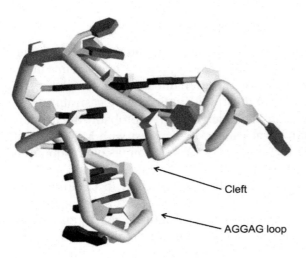

Cleft

AGGAG loop

FIGURE 7–10 One of the ensemble of NMR structures for the *c-kit1* quadruplex (Phan *et al.*, 2007).

two single-nucleotide propeller loops, a lateral-type two-nucleotide loop and a novel large five-nucleotide loop which terminates in the guanine insertion (into the central G-quartet of the stack). The large AGGAG loop forms a cleft in the structure which has the dimensions to be a plausible small-molecule binding site (Figure 7–10), although this cleft is observed to be smaller in the crystal structure (Wei *et al.*, to be published). The 21-mer *c-kit2* quadruplex structures are more complex (albeit with less unusual features) in that the native sequence does not form a single species in solution (Hsu *et al.*, 2009; Kuryavyi, Phan & Patel, 2010), and it was necessary to mutate the G21 residue to T21 in order to obtain high-quality NMR spectra. There appears to be more than one closely related monomeric form (Figure 7–11a), all with parallel topologies (Hsu *et al.*, 2009; Kuryavyi, Phan & Patel, 2010). They have three propeller loops, two single-nucleotide ones (C and A), together with a long (CGCGA) one. In addition a novel all-parallel dimeric *c-kit2* quadruplex form (Figure 7–11b) has been reported (Kuryavyi, Phan & Patel, 2010), which may be relevant to recombination processes.

A number of small-molecule ligands have been examined for *c-kit* quadruplex binding, and for their effects on KIT expression in cells. It is common for significant differences in affinity between the two *c-kit* quadruplexes to be found, which undoubtedly reflect their differences in tertiary structure and likely binding sites, although as yet there are no structural data available on any of the *c-kit* ligand complexes. In common with the *c-myc* ligands, the *c-kit* ones described to date are structurally diverse, although almost all contain a polycyclic aromatic core platform (Figure 7–12). The tetrasubstituted naphthalene diimide compound (Figure 7–12a) stabilizes the *c-kit2* quadruplex to a greater extent than the *c-kit1* one, with exceptionally high ΔT_m values of 29.0 and 11.2° at a 0.5 μM ligand concentration (measured by a FRET melting method: Gunaratnam *et al.*, 2009). By comparison, Gleevec itself has an insignificant effect on thermal stability

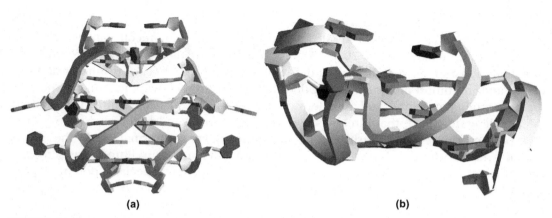

(a)   (b)

FIGURE 7–11 The two types of quadruplex formed by the *c-kit2* quadruplex, as determined by NMR methods. **(a)** shows one of the ensemble of *c-kit2* dimer structures (Kuryavyi, Phan & Patel, 2010), and **(b)** shows a *c-kit2* monomeric quadruplex structure (Hsu *et al.*, 2009).

with ΔT_m values of 0.2° for both the *c-kit1* and *c-kit2* quadruplexes. The established quadruplex-binding ligands BRACO-19 (Chapter 4) and the porphyrin TMPyP4 gave ΔT_m values which were considerably less than that of the naphthalene diimide, but both favored the *c-kit2* quadruplex. Explanations for this preference await detailed structural data, although it might be that these ligands are too large to bind in the *c-kit1* quadruplex cleft and instead have a preference for the more open *c-kit2* structures. A series of trisubstituted isoalloxazine compounds (Figure 7–12b), on the other hand, show a consistent preference for binding to the *c-kit1* quadruplex compared to the *c-kit2* one (Bejugam *et al.*, 2007), with the best compound in this series giving ΔT_m values of 27.1 vs. 17.2°. Similar patterns of preferences are shown by a family of 6-monosubstituted indenoisoquinolines (Figure 7–12d: Bejugam *et al.*, 2010) and by a series of disubstituted diethynylpyridine amides (Figure 7–12e: Dash *et al.*, 2011). Conversely, a series of monosubstituted benzo[a]phenoxazines (Figure 7–12c: McLuckie *et al.*, 2011) show a significant preference for the *c-kit2* quadruplex, by 7.5- to 9.6-fold. A distinct approach to selectively targeting the *c-kit1* quadruplex has utilized a family of five short peptide nucleic acid (PNA) sequences as probes (Amato *et al.*, 2011) that could bind to the long AGGAG loop. A range of biophysical methods were used to monitor interactions, together with molecular modeling, and indicated that short sequences such as (NH_3-lysine-$T^PC^PC^PT^PC^P$, where X^P indicates a PNA) can bind with high affinity and do not disrupt the quadruplex.

Several of these small-molecule ligands have been examined for their cellular effects on *c-kit* expression. The naphthalene diimide compound, for example, produces dose-dependent reductions in mRNA levels in the GIST882 patient-derived primary cancer cell line (Gunaratnam *et al.*, 2009), with a ~90% reduction at the IC_{50} concentration of ~1 μM, over a 24 h period: the other G-quadruplex ligands examined in this study, BRACO-19 and TMPyP4, had no effect. An isoalloxazine compound (Figure 7–12b) produced analogous effects in the HGC-27 gastric carcinoma cell line (Bejugam *et al.*, 2007), as did the indenoisoquinolines in both the GIST882 and the HT-29 colorectal cancer cell lines (Figure 7–12d: Bejugam *et al.*, 2010). All of these data provide correlative evidence of a direct relationship between ligand binding to and stabilizing the *c-kit* promoter quadruplexes and inhibition of *c-kit* transcription, but are not direct evidence linking the two. A recent study has established a luciferase reporter assay, transfected into HGC-27 cells (McLuckie *et al.*, 2011), that directly shows the effects of ligand-induced changes at the *c-kit* promoter on *c-kit* transcription. Crucially the effects are normalized to the effects of a second reporter that is quadruplex-negative. A 173 compound library was screened with this assay and two hits were found, of which one was especially potent (Figure 7–12d). It is not known whether the quadruplex binding preference for *c-kit2* over *c-kit1* plays a role in the biological potency of these and the naphthalene diimide compound

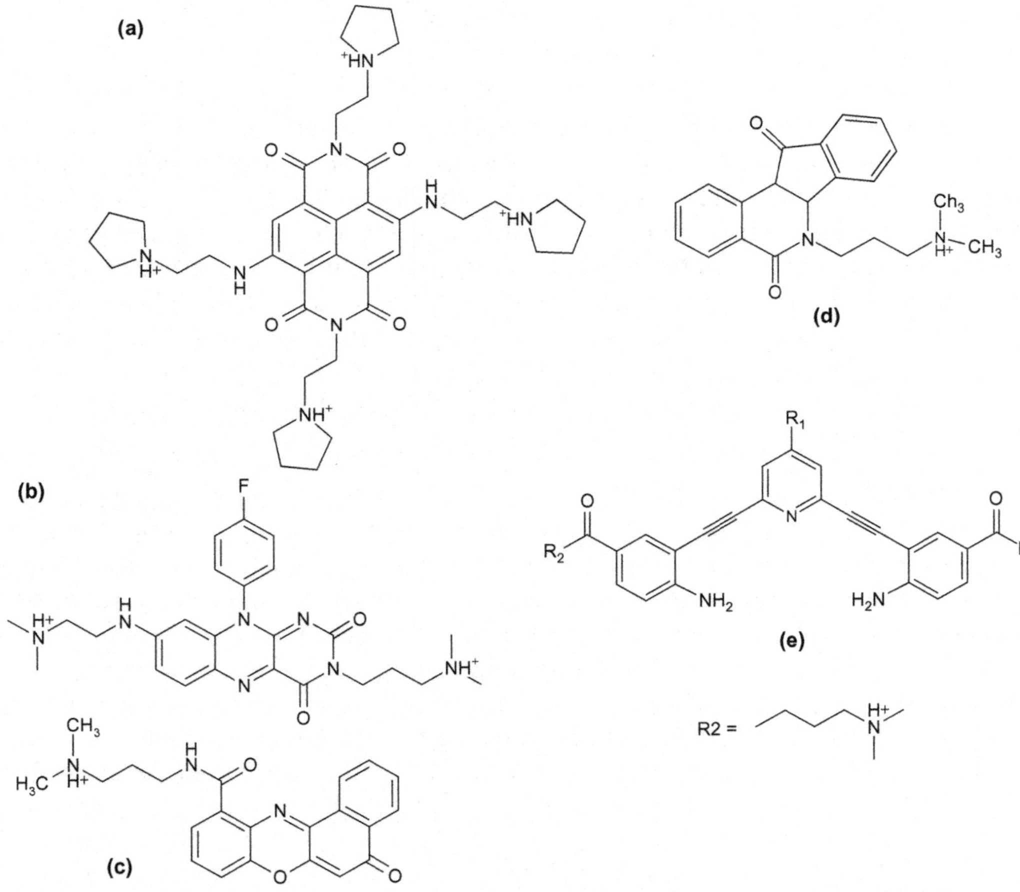

FIGURE 7–12 Small-molecule ligands that have been reported to bind to *c-kit* quadruplexes. See the text for details.

described here, but at this stage both compound classes appear to have potential for further study in GIST cells and tumors.

Other Genomic Quadruplexes

A number of other oncogene promoter sequences have now been identified. In the case of the *bcl-2* promoter, an NMR structure has been determined of a slightly mutated sequence (Dai *et al.*, 2006) and shown to be a mixed parallel/anti-parallel quadruplex with two lateral loops and one propeller loop (Figure 7–13). One of these lateral loops (AGGAATT) is exceptionally long and relatively flexible, at least in the absence of a ligand. The complementary strand forms a relatively stable i-motif structure (Kendrick *et al.*, 2009). The 23-mer G-quadruplex sequence studied by NMR formed part of a longer (39-mer) sequence in the *bcl-2* promoter that has six G-tracts and plays a significant role in the transcription of this gene. A number of small-molecule ligands based on quindoline compounds monosubstituted with an alkylamino side-chain have been examined as inhibitors of *bcl-2* transcription (Wang *et al.*, 2010) using a cell-based *bcl-2* transcription reporter assay. Evidence for the involvement of the promoter quadruplex comes from experiments in which the quadruplex sequence is mutated so that its structure cannot form; this results in a reduction in *bcl-2* transcription. Since this gene encodes a protein that is over-expressed and blocks apoptosis in a number of cancers, especially following the onset of resistance to chemotherapeutic agents, it is itself an important therapeutic target and the development of more selective and potent ligands is warranted.

The *k-ras* oncogene is implicated in a significant proportion of human pancreatic cancers. The evolutionarily conserved promoter region of the human *k-ras* gene contains a nuclease hypersensitive G-rich sequence between nucleotides −327 and −296 upstream from the transcription start site, which has been identified as a putative

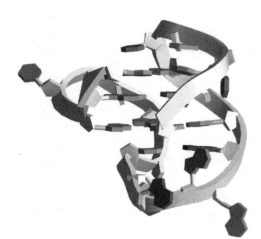

FIGURE 7–13 Cartoon representation of one of the ensemble of NMR structures for the *bcl-2* quadruplex (Dai *et al.*, 2006).

quadruplex-forming region (Cogio & Xodo, 2006; Cogio et al., 2008; Paramasivan, Cogio & Xodo, 2011) and experimentally validated as quadruplex forming by a panel of biophysical and chemical probe methods, although molecular structures have not yet been determined by crystallography or NMR. Several proteins have also been identified as binding to the *k-ras* quadruplexes, including poly[ADP-ribose] polymerase I (PARP-1) (see also Chapter 1) and the myc-associated zinc finger protein MAZ (Cogio et al., 2010), suggesting that they, together with the *k-ras* quadruplexes, are involved in the transcription of this gene. One series of small molecules, guanidine phthalocyanines, have been reported (Membrino *et al.*, 2010) as binding strongly to *k-ras* quadruplexes and inhibiting transcription of this gene, using a reporter assay approach. Whether these molecules have sufficient cellular selectivity for this target remains to be ascertained.

Other promoter sequences demonstrated to have quadruplex formation include the hTERT promoter, where a 20-mer fragment (from a much longer ~90-nucleotide run of 12 G-tracts) has been shown to form two distinct quadruplexes, one parallel and one a (3 + 1) quadruplex by NMR methods (Lim *et al.*, 2010). Under molecular crowding conditions the equilibrium moves in favor of the former structure. An end-to-end stacked structure has been proposed (Palumbo, Ebbinghaus & Hurley, 2009) for the full-length sequence, incorporating the equivalent parallel quadruplex linked to an unusual quadruplex with a 26-nucleotide loop, on the basis of chemical protection and biophysical studies. Promoter regions for the growth factors PDGF (Platelet-Derived Growth Factor) and the VEGF (Vascular Endothelial Growth Factor) have been identified (Qin *et al.*, 2007; Sun, Guo & Shin, 2010), and transcription of the PDGF gene shown to be down-regulated by telomestatin (Qin *et al.*, 2010). The therapeutically important HIF gene is up-regulated in a number of tumor types where it is associated with hypoxia; again a quadruplex-forming sequence has been identified and validated *in vitro* (De Armond *et al.*, 2005). Surprisingly no small-molecule ligands have as yet been described that down-regulate its transcription.

This list of known promoter quadruplex targets is not exhaustive, and undoubtedly many more remain to be discovered. The challenges for the future are (i) to understand the role of these structures in the natural transcriptional processes of these genes, (ii) to determine the nature of the qaudruplexes in detail, including their structures, and (iii) to use this information in the design and development of specific small-molecule regulators of quadruplex formation and transcription.

References

Agarwal, T., Roy, S., Chakraborty, T.K., Maiti, S., 2010. Selective targeting of G-quadruplex using furan-based cyclic homooligopeptides: effect on c-MYC expression. Biochemistry 49, 8388–8397.

Amato, J., Pagano, B., Borbone, N., Oliviero, G., Gabelica, V., De Pauw, E., et al., 2011. Targeting G-quadruplex structure in the human c-kit promoter with short PNA sequences. Bioconjug. Chem. In press.

Ambrus, A., Chen, D., Dai, J., Jones, R.A., Yang, D., 2005. Solution structure of the biologically relevant G-quadruplex element in the human c-MYC promoter. Implications for G-quadruplex stabilization. Biochemistry 44, 2048–2058.

Balasubramanian, S., Hurley, L.H., Neidle, S., 2011. Targeting G-quadruplexes in gene promoters: a novel anticancer strategy? Nature Rev. Drug Discov. 10, 261–275.

Bejugam, M., Gunaratnam, M., Müller, S., Sanders, D.A., Sewitz, S., Fletcher, J., et al., 2010. Targeting the c-Kit promoter quadruplexes with 6-substituted indenoisoquinolines, a novel class of G-quadruplex stabilising small molecule ligands. ACS Med. Chem. Lett. 1, 306–310.

Bejugam, M., Sewitz, S., Shirude, P.S., Rodriguez, R., Shahid, R., Balasubramanian, S., 2007. Trisubstituted isoalloxazines as a new class of G-quadruplex binding ligands: small molecule regulation of c-kit oncogene expression. J. Amer. Chem. Soc. 129, 12926–12927.

Brooks, T.A., Hurley, L.H., 2009. The role of supercoiling in transcriptional control of MYC and its importance in molecular therapeutics. Nature Rev. Cancer 9, 849–861.

Brooks, T.A., Hurley, L.H., 2010. Targeting MYC expression through G-quadruplexes. Genes Cancer 1, 641–649.

Brooks, T.A., Kendrick, S., Hurley, L.H., 2010. Making sense of G-quadruplex and i-motif functions in oncogene promoters. FEBS J. 277, 3459–3469.

Capra, J.A., Paeschke, K., Singh, M., Zakian, V.A., 2010. G-quadruplex DNA sequences are evolutionarily conserved and associated with distinct genomic features in Saccharomyces cerevisiae. PLoS Comput. Biol. 6, e1000861.

Chan, D.S., Yang, H., Kwan, M.H., Cheng, Z., Lee, P., Bai, L.-P., et al., 2011. Structure-based optimization of FDA-approved drug methylene blue as a c-myc G-quadruplex DNA stabilizer. Biochimie 93, 1055–1064.

Cogoi, S., Xodo, L.E., 2006. G-quadruplex formation within the promoter of the KRAS proto-oncogene and its effect on transcription. Nucleic Acids Res. 34, 2536–2549.

Cogoi, S., Paramasivam, M., Membrino, A., Yokoyama, K.K., Xodo, L.E., 2010. The KRAS promoter responds to Myc-associated zinc finger and poly(ADP-ribose) polymerase 1 proteins, which recognize a critical quadruplex-forming GA-element. J. Biol. Chem. 285, 22003–22016.

Cogoi, S., Paramasivam, M., Spolaore, B., Xodo, L.E., 2008. Structural polymorphism within a regulatory element of the human KRAS promoter: formation of G4-DNA recognized by nuclear proteins. Nucleic Acids Res. 36, 3765–3780.

Dai, J., Ambrus, A., Hurley, L.H., Yang, D., 2009. A direct and nondestructive approach to determine the folding structure of the i-motif DNA secondary structure by NMR. J. Amer. Chem. Soc. 131, 6102–6104.

Dai, J., Chen, D., Jones, R.A., Hurley, L.H., Yang, D., 2006. NMR solution structure of the major G-quadruplex structure formed in the human BCL-2 promoter region. Nucleic Acids Res. 34, 5133–5144.

Dang, C.V., 2010. Enigmatic MYC conducts an unfolding systems biology symphony. Genes Cancer 1, 526–531.

Dash, J., Waller, Z.A., Panto, G.D., Balasubramanian, S., 2011. Synthesis and binding studies of novel diethynyl-pyridine amides with genomic promoter DNA G-quadruplexes. Chemistry 17, 4571–4581.

De Armond, R., Wood, S., Sun, D., Hurley, L.H., Ebbinghaus, S.W., 2005. Evidence for the presence of a guanine quadruplex forming region within a polypurine tract of the hypoxia inducible factor 1α promoter. Biochemistry 44, 16341–16350.

Dexheimer, T.S., Carey, S.S., Zuohe, S., Gokhale, V.M., Hu, X., Murata, L.B., et al., 2009. NM23-H2 may play an indirect role in transcriptional activation of c-myc gene expression but does not cleave the nuclease hypersensitive element III1. Mol. Cancer Ther. 8, 1363–1377.

Du, Z., Zhao, Y., Li, N., 2009. Genome-wide colonization of gene regulatory elements by G4 DNA motifs. Nucleic Acids Res. 37, 6784–6798.

Eddy, J., Maizels, N., 2008. Conserved elements with potential to form polymorphic G-quadruplex structures in the first intron of human genes. Nucleic Acids Res. 36, 1321–1333.

Eddy, J., Vallur, A.C., Varma, S., Liu, H., Reinhold, W.C., Pommier, Y., et al., 2011. G4 motifs correlate with promoter-proximal transcriptional pausing in human genes. Nucleic Acids Res. In press.

Fernando, H., Reszka, A.P., Huppert, J., Ladame, S., Rankin, S., Venkitaraman, A.R., et al., 2006. A conserved quadruplex motif located in a transcription activation site of the human c-kit oncogene. Biochemistry 45, 7854–7860.

Fletcher, J.A., Rubin, B.P., 2007. KIT mutations in GIST. Curr. Opin. Genet. Dev. 7, 3–7.

González, V., Hurley, L.H., 2010. The C-terminus of nucleolin promotes the formation of the c-MYC G-quadruplex and inhibits c-MYC promoter activity. Biochemistry 49, 9706–9714.

González, V., Hurley, L.H., 2010. The c-MYC NHE III(1): function and regulation. Ann. Rev. Pharmacol. Toxicol. 50, 111–129.

Grand, C.L., Han, H., Munoz, R.M., Weitman, S., Von Hoff, D.D., Hurley, L.H., et al. 2002. The cationic porphyrin TMPyP4 down-regulates c-MYC and human telomerase reverse transcriptase expression and inhibits tumor growth in vivo. Mol. Cancer Ther. 1, 565–573.

Gunaratnam, M., Swank, S., Haider, S.M., Galesa, K., Reszka, A.P., Beltran, M., et al., 2009. Targeting human gastrointestinal stromal tumor cells with a quadruplex-binding small molecule. J. Med. Chem. 52, 3774–3783.

Hatzakis, E., Okamoto, K., Yang, D., 2010. Thermodynamic stability and folding kinetics of the major G-quadruplex and its loop isomers formed in the nuclease hypersensitive element in the human c-Myc promoter: effect of loops and flanking segments on the stability of parallel-stranded intramolecular G-quadruplexes. Biochemistry 49, 9152–9160.

Hershman, S.G., Chen, Q., Lee, J.Y., Kozak, M.L., Yue, P., Wang, L.S., et al., 2008. Genomic distribution and functional analyses of potential G-quadruplex-forming sequences in Saccharomyces cerevisiae. Nucleic Acids Res. 36, 144–156.

Hsu, S.-T., Varnai, P., Bugaut, A., Reszka, A.P., Neidle, S., Balasubramanian, S., 2009. A G-rich sequence within the c-kit oncogene promoter forms a parallel G-quadruplex having asymmetric G-tetrad dynamics. J. Amer. Chem. Soc. 131, 13399–133409.

Huppert, J.L., Balasubramanian, S., 2007. G-quadruplexes in promoters throughout the human genome. Nucleic Acids Res. 35, 406–413.

Hurley, L.H., Von Hoff, D.D., Siddiqui-Jain, A., Yang, D., 2006. Drug targeting of the c-MYC promoter to repress gene expression via a G-quadruplex silencer element. Semin. Oncol. 33, 498–512.

Kendrick, S., Akiyama, Y., Hecht, S.M., Hurley, L.H., 2009. The i-motif in the bcl-2 P1 promoter forms an unexpectedly stable structure with a unique 8:5:7 loop folding pattern. J. Amer. Chem. Soc. 131, 17667–17676.

Kuryavyi, V., Phan, A.T., Patel, D.J., 2010. Solution structures of all parallel-stranded monomeric and dimeric G-quadruplex scaffolds of the human c-kit2 promoter. Nucleic Acids Res. 38, 6757–6773.

Lim, K.W., Lacroix, L., Yue, D.J., Lim, J.K., Lim, J.M., Phan, A.T., 2010. Coexistence of two distinct G-quadruplex conformations in the hTERT promoter. J. Amer. Chem. Soc. 132, 12331–12342.

Ma, Y., Ou, T.-M., Tan, J.-H., Hou, J.-Q., Huang, S.-L, Gu, L.-Q., et al., 2011. Quinolino-benzo-[5,6]-dihydroisoquindolium compounds derived from berberine: a new class of highly selective ligands for G-quadruplex DNA in c-myc oncogene. Eur. J. Med. Chem. 46, 1906–1913.

McLuckie, K.I., Waller, Z.A., Sanders, D.A., Alves, D., Rodriguez, R., Dash, J., et al., 2011. G-quadruplex-binding benzo[a]phenoxazines down-regulate c-KIT expression in human gastric carcinoma cells. J. Amer. Chem. Soc. 133, 2658–2663.

Membrino, A., Paramasivam, M., Cogoi, S., Alzeer, J., Luedtke, N.W., Xodo, L.E., 2010. Cellular uptake and binding of guanidine-modified phthalocyanines to KRAS/HRAS G-quadruplexes. Chem. Commun. (Camb.) 46, 625–627.

Murchie, A.I., Lilley, D.M., 1992. Retinoblastoma susceptibility genes contain 5′ sequences with a high propensity to form guanine-tetrad structures. Nucleic Acids Res. 11, 49–53.

Palumbo, S.M., Ebbinghaus, S.W., Hurley, L.H., 2009. Formation of a unique end-to-end stacked pair of G-quadruplexes in the hTERT core promoter with implications for inhibition of telomerase by G-quadruplex-interactive ligands. J. Amer. Chem. Soc. 131, 10878–10891.

Paramasivam, M., Cogoi, S., Xodo, L.E., 2011. Primer extension reactions as a tool to uncover folding motifs within complex G-rich sequences: analysis of the human KRAS NHE. Chem. Commun. (Camb.) In press

Parkinson, G.N., Lee, M.P.H., Neidle, S., 2002. Crystal structure of parallel quadruplexes from human telomeric DNA. Nature 417, 881–884.

Pelengaris, S., Khan, M., Evan, G., 2002. c-MYC: more than just a matter of life and death. Nature Rev. Cancer 2, 764–776.

Peng, D., Tan, J.-H., Chen, S.-B., Ou, T.-M., Gu, L.-Q., Huang, Z.-S., 2010. Bisaryldiketene derivatives: a new class of selective ligands for c-myc G-quadruplex DNA. Bioorg. Med. Chem. 18, 8235–8242.

Phan, A.T., Mergny, J.-L., 2002. Human telomeric DNA: G-quadruplex, i-motif and Watson-Crick double helix. Nucleic Acids Res. 30, 4618–4625.

Phan, A.T., Guéron, M., Leroy, J.-L., 2000. The solution structure and internal motions of a fragment of the cytidine-rich strand of the human telomere. J. Mol. Biol. 299, 123–144.

Phan, A.T., Kuryavyi, V., Burge, S., Neidle, S., Patel, D.J., 2007. Structure of an unprecedented G-quadruplex scaffold in the human c-kit promoter. J. Amer. Chem. Soc. 129, 4386–4392.

Phan, A.T., Kuryavyi, V., Gaw, H.Y., Patel, D.J., 2005. Small-molecule interaction with a five-guanine-tract G-quadruplex structure from the human MYC promoter. Nature Chem. Biol. 1, 167–173.

Phan, A.T., Modi, Y.S., Patel, D.J., 2004. Propeller-type parallel-stranded G-quadruplexes in the human c-myc promoter. J. Amer. Chem. Soc. 126, 8710–8716.

Qin, Y., Fortin, J.S., Tye, D., Gleason-Guzman, M., Brooks, T.A., Hurley, L.H., 2010. Molecular cloning of the human platelet-derived growth factor receptor beta (PDGFR-beta) promoter and drug targeting of the G-quadruplex-forming region to repress PDGFR-beta expression. Biochemistry 49, 4208–4219.

Qin, Y., Rezler, E.M., Gokhale, V., Sun, D., Hurley, L.H., 2007. Characterization of the G-quadruplexes in the duplex nuclease hypersensitive element of the PDGF-A promoter and modulation of PDGF-A promoter activity by TMPyP4. Nucleic Acids Res. 35, 7698–7713.

Rangan, A., Fedoroff, O.Y., Hurley, L.H., 2001. Induction of duplex to G-quadruplex transition in the c-myc promoter region by a small molecule. J. Biol. Chem. 276, 4640–4646.

Rankin, S, Reszka, A.P., Huppert, J., Zloh, M., Parkinson, G.N., Todd, A.K., et al., 2005. Putative DNA quadruplex formation within the human c-kit oncogene. J. Amer. Chem. Soc. 127, 10584–10589.

Rawal, P., Kummarasetti, V.B., Ravindran, J., Kumar, N., Halder, K., Sharma, R., et al., 2006. Genome-wide prediction of G4 DNA as regulatory motifs: role in Escherichia coli global regulation. Genome Res. 16, 644–655.

Ren, J., Chaires, J.B., 1999. Sequence and structural selectivity of nucleic acid binding ligands. Biochemistry 38, 16067–16075.

Shalaby, T., von Bueren, A.O., Hürlimann, M.L., Fiaschetti, G., Castelletti, D., Masayuki, T., et al., 2010. Disabling c-Myc in childhood medulloblastoma and atypical teratoid/rhabdoid tumor cells by the potent G-quadruplex interactive agent S2T1-6OTD. Mol. Cancer Ther. 9, 167–179.

Shirude, P.S., Okumus, B., Ying, L., Ha, T., Balasubramanian, S., 2007. Single-molecule conformational analysis of G-quadruplex formation in the promoter DNA duplex of the proto-oncogene c-kit. J. Amer. Chem. Soc. 129, 7484–7485.

Shirude, P.S., Ying, L., Balasubramanian, S., 2008. Single-molecule conformational analysis of the biologically relevant DNA G-quadruplex in the promoter of the proto-oncogene c-MYC. Chem. Comm. (Camb.) 7, 2007–2009.

Siddiqui-Jain, A., Grand, C.L., Bearss, D.J., Hurley, L.H., 2002. Direct evidence for a G-quadruplex in a promoter region and its targeting with a small molecule to repress c-MYC transcription. Proc. Natl. Acad. Sci. USA 99, 11593–11598.

Simonsson, T., 2001. G-quadruplex DNA structures—variations on a theme. Biol. Chem. 382, 621–628.

Simonsson, T., Pecinka, P., Kubista, M., 1998. DNA tetraplex formation in the control region of c-myc. Nucleic Acids Res. 26, 1167–1172.

Simonsson, T., Pribylova, M., Vorlickova, M., 1999. A nuclease hypersensitive element in the human c-myc promoter adopts several distinct i-tetraplex structures. Biochem. Biophys. Res. Commun. 278, 158–168.

Sun, D., Hurley, L.H., 2009. The importance of negative superhelicity in inducing the formation of G-quadruplex and i-motif structures in the c-Myc promoter: implications for drug targeting and control of gene expression. J. Med. Chem. 52, 2863–2874.

Sun, D., Guo, K., Shin, Y.J., 2011. Evidence of the formation of G-quadruplex structures in the promoter region of the human vascular endothelial growth factor gene. Nucleic Acids Res. 39, 1256–1265.

Todd, A.K., Haider, S.M., Parkinson, G.N., Neidle, S., 2007. Sequence occurrence and structural uniqueness of a G-quadruplex in the human c-kit promoter. Nucleic Acids Res. 35, 5799–5808.

Todd, A.K., Neidle, S., 2008. The relationship of potential G-quadruplex sequences in cis-upstream regions of the human genome to SP1-binding elements. Nucleic Acid Res. 36, 2700–2704.

Tuveson, D.A., Willis, N.A., Jacks, T., Griffin, J.D., Singer, S., Fletcher, C.D., et al., 2001. STI571 inactivation of the gastrointestinal stromal tumor c-KIT oncoprotein: biological and clinical implications. Oncogene 20, 5054–5058.

Verma, A., Halder, K., Halder, R., Yadav, V.K., Rawal, P., Thakur, R.K., et al., 2008. Genome-wide computational and expression analyses reveal G-quadruplex DNA motifs as conserved cis-regulatory elements in human and related species. J. Med. Chem. 51, 5641–5649.

Verma, A., Yadav, V.K., Basundra, R., Kumar, A., Chowdhury, S., 2009. Evidence of genome-wide G4 DNA-mediated gene expression in human cancer cells. Nucleic Acids Res. 37, 4194–4204.

Wang, P., Leung, C.-H., Ma, D.-L., Yan, S.-C., Che, C.M., 2010. Structure-based design of platinum(II) complexes as c-myc oncogene down-regulators and luminescent probes for G-quadruplex DNA. Chemistry 16, 6900–6911.

Wang, X.-D., Ou, T.-M., Lu, Y.-J, Li, Z, Xu, Z., Xi, C., et al., 2010. Turning off transcription of the bcl-2 gene by stabilizing the bcl-2 promoter quadruplex with quindoline derivatives. J. Med. Chem. 53, 4390–4398.

Xu, Y., Sugiyama, H., 2006. Formation of the G-quadruplex and i-motif structures in retinoblastoma susceptibility genes (Rb). Nucleic Acids Res. 34, 949–954.

Introduction

The concept that RNA nucleic acids can also form four-stranded structures is not recent—the first four-stranded guanine-containing helix to be characterized was the polynucleotide poly (rG). Biophysical studies on a number of RNA quadruplexes have shown that they are typically very stable, sometimes more so than their DNA counterparts (Mergny *et al.*, 2005). Many if not all examined to date appear to adopt a singular parallel topology in dilute solution, contrasting with the polymorphic diversity shown by many DNA quadruplexes (Joachimi, Benz & Hartig, 2009), whatever the environmental conditions such as changes in ionic environment from Na^+ to K^+ (Zhang *et al.*, 2010; Zhang & Zhi, 2010).

The attraction of RNA quadruplexes as therapeutic targets stems in large part from the view that RNA targets are derived from transcribed genes, so will be inherently single stranded (at least in principle), in contrast to genomic DNA quadruplexes, which have to be formed from genomic duplex DNA, and thus either require motive force to be produced or need to be captured during replication. RNA quadruplexes from 5′-UTR (untranslated) RNA sequences have been the focus of particular interest and are described in Section 8.3. Quadruplexes from telomeric RNAs have very recently emerged as potential targets, and these are also described below.

Very few 3-D structures of RNA quadruplexes have been determined to date, undoubtedly in part because of the high cost (and low yields relative to DNA sequences) of milligram-scale RNA synthesis and purification. Since both crystallography and NMR need to be able to screen a number of sequence variants (see Chapter 10) in order to obtain either diffraction-quality crystals or NMR spectra corresponding to single species, the barrier to their widespread study remains high.

Telomeric RNA Quadruplexes

Telomeric DNA has always been considered to be transcriptionally inert, the "end of the line." However, two reports very surprisingly demonstrated that this is not the case and that telomeres are frequently transcribed in many organisms by RNA polymerase II (Luke & Lingner, 2009; Schoeftner & Blasco, 2009) into a large number of relatively shorter telomeric RNA transcripts (Azzalin *et al.*, 2007; Schoefner & Blasco, 2008), termed TERRA sequences (Telomeric Repeat containing RNA). The majority of human TERRA molecules are ~200

Therapeutic Applications of Quadruplex Nucleic Acids. DOI: 10.1016/B978-0-12-375138-6.00008-X

139

Table 8–1 Quadruplex RNA Structures Available from the PDB

Sequence	Method	Reference	PDB id
r(UGGGGU)	NMR	Cheong & Moore, 1992	1RAU
r(UAGGGUUAGGGU)	NMR	Martadinata & Phan, 2009	2KBP
r(U^{Br}AGGGUUAGGGU)	X-ray	Collie et al., 2010	3BIK
r(UAGGGUUAGGGU)	X-ray	Collie et al., 2011	3MIJ

nucleotides long (Porro et al., 2010), and are associated with a large number of RNA-binding and chromatin-associated proteins (López de Silanes, d'Alcontres & Blasco, 2010), as well as with the telomere-repeat protein TRF2 (Deng et al., 2009), and may play a role in telomere maintenance. Although the central role of TERRA (if it has one) is not yet clear, it is evident that it is involved in several important telomere-related pathways: for example, it appears to be a direct inhibitor of telomerase (Redon, Reichenbach & Lingner, 2010), perhaps by binding to the RNA template.

In view of the evidence that TERRA molecules are associated with a number of proteins, it is interesting that TERRA molecules have been localized in a quadruplex form within cell nuclei (Xu et al., 2010). This is perhaps unsurprising in view of the high stability of RNA telomeric quadruplexes (see Section 8.2 below). A study of long TERRA transcripts using circular dichroism and electron microscopy (Randall & Griffith, 2009) has shown that these form compact thickened rods whose dimensions are compatible with the formation of discrete parallel quadruplexes from groups of four UUAGGG repeats. TERRA is unable to release the single-stranded DNA binding protein POT1 from the telomeric overhang, for steric reasons (Nandakumar, Podell & Cech, 2010), and thus does not readily form a DNA–RNA hybrid duplex (or quadruplex) at the overhang. TERRA also actively aids POT1 binding to the overhang by binding directly to and removing the heterogeneous nuclear ribonucleoprotein hnRNPA1 (Flynn et al., 2011). Whether TERRA is a therapeutic target is unclear at present, although specific ligands for TERRA will undoubtedly aid in unraveling its function.

Structural studies on simple tetramolecular RNA quadruplexes (for example, r(UGGGGU): Cheong & Moore, 1992) have indicated that these adopt very similar structures that are closely similar to their deoxy- counterparts. More recently the detailed structure of a bimolecular telomeric RNA quadruplex has been defined by both NMR (Martadinata & Phan, 2009) and crystallography (Collie et al., 2010). Unusually both methods analyzed the same sequence (Table 8–1). Crystallography and NMR concur in demonstrating that this sequence in K$^+$ solution forms a parallel quadruplex (Figure 8–1a, b), which is also in agreement with circular dichroism spectra that show a positive peak at 260 nm and a trough at 240 nm (Xu, Kaminaga & Komiyama, 2008; Arora & Maiti, 2009; Collie et al., 2009; Martadinata & Phan, 2009). This topology is retained for the 12-mer in Na$^+$ solution (Xu, Kaminaga & Komiyama, 2008), and also for four-repeat sequences in K$^+$ solution (Martadinata & Phan, 2009), in contrast with the polymorphic behavior of

(a) **(b)**

FIGURE 8–1 (a) Representation of one of the ensemble of NMR structures of the RNA bimolecular telomeric quadruplex (Martadinata & Phan, 2009). **(b)** Representation of the crystal structure of the RNA bimolecular telomeric quadruplex (Collie *et al.*, 2009). The two K^+ ions in the ion channel are shown.

telomeric DNA quadruplexes in dilute solution (see Chapter 2). The overall features of this quadruplex are closely similar in the two analyses, with three stacked G-quartets forming the core, two UUA propeller loops and all glycosidic angles in the *anti* conformation. Most sugar puckers are C3′-*endo*, although several nucleosides adopt a C2′-*endo* conformation, especially those in the loops. The crystal structure in particular also very closely resembles the original crystallographically determined telomeric DNA bimolecular quadruplex structure; both DNA and RNA crystal structures have a K^+ ion situated midway between each pair of G-quartet planes, coordinated to O6 atoms of guanines in a bipyramidyl antiprismatic manner. The UUA loops in the NMR structure are highly flexible (Figure 8–2), whereas they are well ordered in the crystal structure and have the same overall structure as the TTA loops in the deoxy- telomeric quadruplexes (Parkinson, Lee & Neidle, 2002). The loops in the crystal structure have several short (2.7Å) hydrogen bonds involving O2′ ribo sugar atoms (Figure 8–3), suggesting that these are important contributors to RNA loop stability and thus to the strong preference of telomeric RNA quadruplexes to have a parallel topology. It is notable that individual loop sugar puckers differ between the RNA and DNA quadruplexes, indicating that these are a response to the O2′ hydrogen bonding in the former.

Few studies to date have examined ligand binding to RNA quadruplexes (De Cian *et al.*, 2008; Collie *et al.*, 2009; Lacroix, Séosse & Mergny, 2011). The bis-quinolinium ligand 360A (Figure 8–4a: see also Chapter 5) shows only limited discrimination between DNA and RNA quadruplexes, with a ~7° lower ΔT_m value for the latter using a FRET melting technique at $1\mu M$ ligand concentration. Similar small differences were found with BRACO-19 and several naphthalene diimides (Collie *et al.*, 2009). These differences in ΔT_m values are indicative of reduced stability of RNA quadruplex complexes and may be due to the presence of the additional O2′ groups in the grooves/loops, which restrict the space available for ligand side-chains. The bis-triazole acridine compound (Sparapani *et al.*, 2010) (Figure 8–4b), has an extended length of ~33Å with the triazoles groups coplanar with the central acridine, suggesting that this three-group system could extend

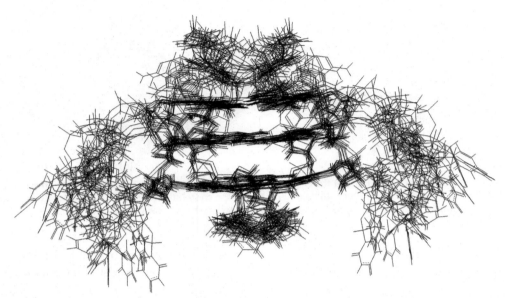

FIGURE 8–2 Superposition of the ensemble of ten NMR structures for the RNA bimolecular telomeric quadruplex (Martadinata & Phan, 2009). Note the increased large spread of conformers for the UUA loops at each side of the structures.

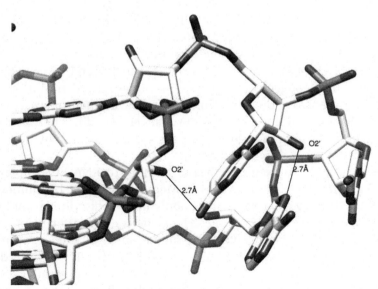

FIGURE 8–3 View of one of the UUA loops in the crystal structure of the RNA bimolecular telomeric quadruplex (Collie *et al.*, 2009), showing the hydrogen bonding involving base…O2′ atoms.

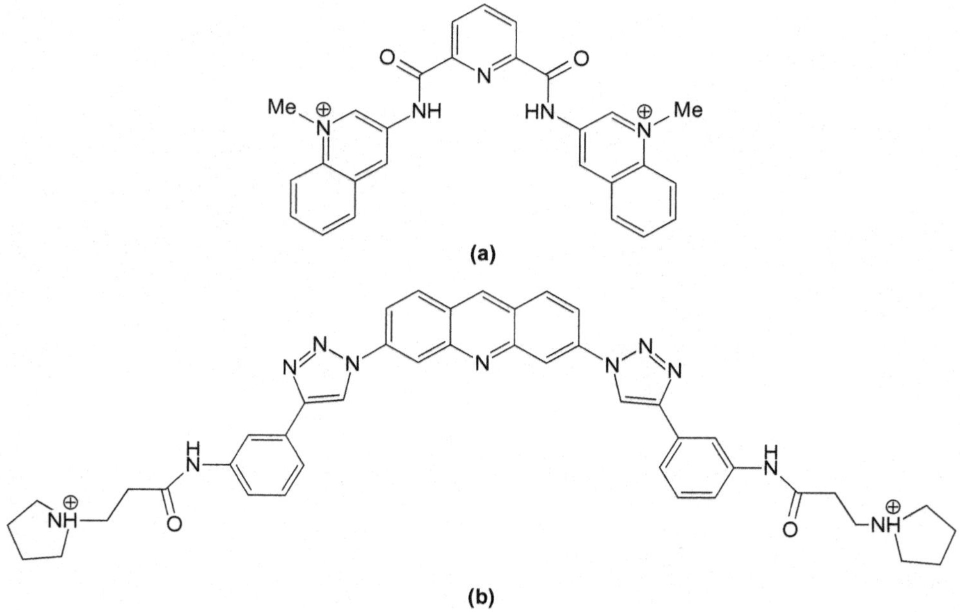

(a)

(b)

FIGURE 8–4 **(a)** The 360A ligand (De Cian *et al.*, 2008). **(b)** The bis-triazole acridine ligand (Sparapani *et al.*, 2010).

beyond a G-quartet. The crystal structure (Collie *et al.*, 2011) of its complex with a bimolecular human telomeric quadruplex shows that two molecules of this ligand bind onto a single G-quartet surface (Figure 8–5a, b). One triazole ring from each molecule also stacks onto an adenine from a UUA loop, which has undergone major conformational changes, stabilized by intramolecular O2′ hydrogen bonding, in order to produce the extended AG4A hexabase binding platform. Such an arrangement is not available when binding to a DNA quadruplex (Figure 8–6), and suggests ways to design ligands selective for RNA vs. DNA telomeric quadruplexes (or vice versa):

1. For optimal discrimination in favor of an RNA telomeric quadruplex, ligands should have minimally sized groove/loop-binding substituents. Alternatively, a ligand molecule can be extended in one dimension to induce the stabilization of additional loop-binding platforms.
2. DNA telomeric quadruplexes will be favored by the majority of ligands developed to date, and increasing the size of substituents and their end-groups to the maximum possible will enhance grove/loop discrimination.

As well as discriminating between DNA and RNA telomeric quadruplexes, such ligands should not bind to any other type of nucleic acid structure, especially DNA or RNA duplexes, hairpins or pseudoknots (which will themselves have distinct morphological

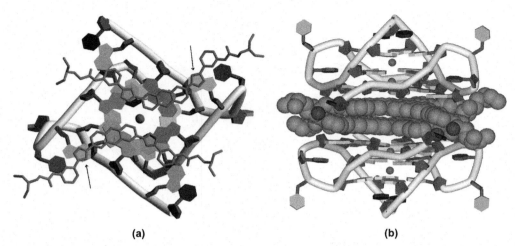

(a)   (b)

FIGURE 8–5 Views of the co-crystal structure between the bis-triazole acridine and the RNA bimolecular telomeric quadruplex (Collie *et al.*, 2011). **(a)** View onto the terminal G-quartet plane. The stacking of a triazole ring, one from each ligand molecule, onto an adenine base from the UUA loop, is indicated by an arrow. This adenine is coplanar with the G-quartet. **(b)** The arrangement in the crystal of two head-to-head quadruplexes, with a total of four ligand molecules (shown in space-filling representation) sandwiched between them.

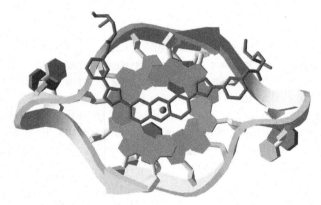

FIGURE 8–6 Cartoon representation of the crystal structure between a diethylamino analog of the bis-triazoles acridine shown in Figure 8–4b, and the DNA telomeric quadruplex d(TAGGGTTAGGGT). A single ligand molecule is shown stacked onto the terminal G-quartet: this is disordered in the crystal around a two-fold axis perpendicular to the page and through the K$^+$ axis, resulting in a second ligand half-molecule. The structure has been reported by Collie *et al.* (2011), and deposited in the PDB with id 3QCR.

features). That this at least can be achieved at the duplex level has been shown (Rzuczek *et al.*, 2010; Satyanarayana *et al.*, 2010) for a series of synthetic macrocycles based on a polyoxazole core that have been inspired by the selectivity of telomestatin. Several of these compounds show high ΔT_m values for the human telomeric quadruplex of up to 24° yet no discernible measured ΔT_m with either duplex DNA or RNA. Their effect on the stability of a

5′-UTR RNA quadruplex was significant with ΔT_m values of up to 34°, although these compounds have not been evaluated with human telomeric RNA quadruplexes. The small size of their substituents, terminating in -NMe$_2$ groups, suggests that this is worth evaluating. A fluorescence-based assay for ligands binding to telomerase RNA quadruplexes vs. other nucleic acids has been described (Lacroix, Séosse & Mergny, 2011), which enables high throughput to be achieved, and in principle is adaptable to other RNA quadruplexes.

RNA Quadruplexes in mRNA

Gene expression in eukaryotics can be regulated at the RNA level through defined sequences in the untranslated regions of genes (UTRs). Bioinformatic tools have been used (Huppert *et al.*, 2008) to survey the 5′-UTR and 3′-UTR sequences of the human genome for putative quadruplex sequences, using the sequence criteria and restraints given in Chapter 1. 5′-UTR sequences with an average length of 243 nucleotides have 4141 potential quadruplex sequences (including both strands in the cDNA sequence) whereas the much larger (average length 899 nucleotides) 3′-UTR sequences have 5041 putative quadruplex sequences. Thus these are significantly less prevalent in 3′-UTR sequences, taking their increased length into account. There is a large clustering of such sequences in the 5′-UTR sequences immediately upstream of the transcription start site, with a rapid fall-off with increasing separation. In functional terms it has been suggested that the presence of a quadruplex in 5′-UTR RNA would hinder transcription initiation as well as translation, and in 3′-UTR RNA sequences would hinder transcription termination (Wieland & Hartig, 2007; Huppert *et al.*, 2008).

The concept of such RNA quadruplexes as regulatory motifs, and as potential targets for therapeutic intervention, is attractive since it is possible that they may be formed *in situ* and be more readily accessible than promoter quadruplexes (Chapter 7). Accordingly there has been considerable activity in characterizing them in a number of genes of therapeutic interest (Table 8–2), and a few studies on the effects of ligands have been

Table 8–2   Some established RNA quadruplexes in 5′-UTR sequences

Gene	Sequence	Possible Topology	Reference
Estrogen receptor α	r(G$_3$UAG$_4$CAAAG$_4$CUG$_4$)	parallel	Derecka *et al.*, 2010
TRF2	r(G$_3$AG$_3$CG$_4$AG$_3$)	parallel	Gomez *et al.*, 2010
BCL-2	r(G$_5$CCGUG$_4$UG$_3$AGCUG$_4$)	parallel	Shahid, Bugaut & Balasubramanian, 2010
VEGF	r(GGAGGAG$_5$AGGAGGA)	parallel	Morris *et al.*, 2010
MT3 MMP	r(GAG$_3$AG$_3$AG$_3$AGAG$_3$A)	parallel	Morris & Basu, 2009
NRAS	r(G$_3$AG$_4$CG$_3$UCUG$_3$)	parallel	Kumari, Bugaut & Balasubramanian, 2008
Zic-1	r(G$_3$UG$_8$CG$_5$AGGCCGG$_4$)	parallel	Arora *et al.*, 2008

reported. As yet there is no detailed structural information (NMR or crystallography) on any of these targets, so rational ligand design is not yet possible. It is notable that typically the sequences are complex with unequal number of guanines in the (presumed) G-tracts, suggesting that the folded arrangements are structurally more complex and diverse than the RNA telomeric quadruplexes. It is also important to bear in mind that large mRNAs can form complex tertiary structures involving such motifs as hairpins and pseudoknots, so that the existence of a putative quadruplex-forming sequence does not necessarily mean that a quadruplex fold will be formed within a complex RNA, even if it is stable as a separate short sequence, since it may be competing with other stable motifs.

The common way to study the effects on transcription of an RNA quadruplex sequence is to construct a plasmid reporter assay that has the target sequence embedded within a longer sequence, ideally the complete 5′-UTR, which would be upstream of a reporter gene (typically firefly luciferase) and downstream of the T7 promoter. Transcription *in vitro* to the corresponding RNA transcript is accomplished using T7 RNA polymerase, followed by translation using rabbit reticulocyte lyase, which can then be quantitatively assessed by measuring changes in the luminescence of luciferase. The employment of several sequences as controls, having a range of mutations in the quadruplex region, is important. In this way it is possible to have confidence in the results from the reporter assay approach as being strongly supportive of the presence of a functional quadruplex fold within the UTR sequence, when taken together with biophysical data on the quadruplex.

Table 8–2 summarizes the majority of current 5′-UTR RNA quadruplex assignments, on the basis of verification from reporter assays coupled with biophysical studies. It is notable that most sequences have unequal numbers of guanines in their G-tracts, even for the apparently simplest of sequences such as that for TRF2 (which is closely similar to the NRAS sequence). Circular dichroism studies (Table 8–2) show that these sequences adopt parallel-type quadruplex folds, which is unsurprising given that most contain single-nucleotide loops (see Chapter 2). The widely prevalent G-tract heterogeneity suggests that some of these RNA quadruplexes may have complex overall structures involving insertion of "loop" guanines into the G-quartet cores, as has been observed for the *c-kit1* promoter quadruplex (see Chapter 7). There may be additional intramolecular interactions arising from the presence of the O2′ atoms, as is common in complex RNA molecules. Stability of these RNA quadruplexes is typically high: the BCL-2 quadruplex has a T_m value of 81° in the presence of 50 mM potassium ion (Shahid, Bugaut & Balasubramanian, 2010); the TRF2 quadruplex could not be melted even at 95°, in the presence of 100 mM K⁺ ion (Gomez *et al.*, 2010).

The effects of a 5′-UTR RNA quadruplex on translation have been demonstrated in the above sequences using reporter assay approaches. Typically significant reductions in translation efficiency have been observed in the presence of the quadruplex motif, for example nearly four-fold for the NRAS quadruplex (Kumari, Bugaut & Balasubramanian, 2008), two-fold for the estrogen receptor α (Derecka *et al.*, 2010), and about 2.5-fold for TRF2 (Gomez *et al.*, 2010). These effects are markedly dependent on the position of the quadruplex motif within the 5'-UTR sequence (Kumari, Bugaut & Balasubramanian, 2008), which suggests that other RNA structural motifs may modulate quadruplex stability and even interact with it. A systematic survey (Beaudoin & Perreault, 2010) of nine potential 5'-UTR quadruplex sequences distributed across diverse gene functional types showed that six can form stable quadruplexes, and that these produced reductions in translation efficiency in the range 1.5–2.5-fold. That these effects are specific for translation and do not affect transcription was shown (Gomez *et al.*, 2010) by the constancy in RNA levels as assessed by RT-PCR. 5'-UTR RNA quadruplex effects have also been demonstrated in human cells, using the reporter plasmids transfected into cells in culture. For example, with the estrogen receptor α quadruplex in an appropriate cell line that expresses the estrogen receptor (Derecka *et al.*, 2010), a 15-fold increase in translation efficiency was observed in the absence of the quadruplex motif. More modest (~2-fold) changes were observed for the BCL-2 5'-UTR quadruplex in several human epithelial and cancer cell lines (Shahid, Bugaut & Balasubramanian, 2010), although as several studies have found, the size of the effects differ between cell lines and may not be apparent in all.

The potential of 5'-UTR RNA quadruplexes as therapeutic targets has started to be studied. The effects of three small-molecule ligands (the pyridine dicarboxamide compound 360A and two bis-quinolinium derivatives) on translation of the TRF2 5'-UTR sequence have been examined *in vitro* with a reporter assay (Gomez *et al.*, 2010), and show large concentration-dependent decreases in translation relative to a mutant sequence. A bis-quinolinium carboxamide derivative, which has some binding selectivity for the NRAS 5'-UTR RNA quadruplex, has been also shown to produce dose-dependent down-regulation of NRAS translation *in vitro* (Bugaut *et al.*, 2010). These data indicate that the approach has potential, and a challenge will be to discover small molecules with high selectivity and low toxicity that can demonstrate targeting a single 5'-UTR RNA quadruplex in disease cells and down-regulate transcription to a sufficient level for therapy. The attraction of these targets is that they are readily amenable to high-throughput screening of compound libraries using the reporter assay format.

References

Arora, A., Maiti, S., 2009. Different biophysical behaviour of human telomeric DNA and RNA quadruplexes. J. Phys. Chem. B 113, 10515–10520.

Arora, A., Dutkiewicz, M., Scaria, V., Hariharan, M., Maiti, S., Kurreck, J., 2008. Inhibition of translation in living eukaryotic cells by an RNA G-quadruplex motif. RNA 14, 1290–1296.

Azzalin, C.M., Reichenbach, P., Khoriauli, L., Giulotto, E., Lingner, J., 2007. Telomeric repeat containing RNA and RNA surveillance factors at mammalian chromosome ends. Science 318, 798–801.

Beaudoin, J.-D., Perreault, J.-P., 2010. 5'-UTR G-quadruplex structures acting as translational repressors. Nucleic Acids Res. 38, 7022–7036.

Bugaut, A., Rodriguez, R., Kumari, S., Hsu, S.T., Balasubramanian, S., 2010. Small molecule-mediated inhibition of translation by targeting a native RNA G-quadruplex. Org. Biomol. Chem. 8, 2771–2776.

Cheong, C., Moore, P.B., 1992. Solution structure of an unusually stable RNA tetraplex containing both G- and U-quartet structures. Biochemistry 31, 8406–8414.

Collie, G.W., Haider, S.M., Neidle, S., Parkinson, G.N., 2010. A crystallographic and modelling study of a human telomeric RNA (TERRA) quadruplex. Nucleic Acids Res. 38, 5569–5580.

Collie, G.W., Reszka, A.P., Haider, S.M., Gabelica, V., Parkinson, G.N., Neidle, S., 2009. Selectivity in small molecule binding to human telomeric RNA and DNA quadruplexes. Chem. Commun. (Camb.), 7482–7484.

Collie, G.W., Sparapani, S., Parkinson, G.N., Neidle, S., 2011. Structural basis of telomeric RNA quadruplex-acridine ligand recognition. J. Amer. Chem. Soc. 133, 2721–2728.

De Cian, A., Gros, J., Guédin, A., Haddi, M., Lyonnais, S., Guittat, L., et al., 2008. DNA and RNA quadruplex ligands. Nucleic Acids Symp. Ser. (Oxf.) 52, 7–8.

Deng, Z., Norseen, J., Wiedmer, A., Riethman, H., Lieberman, P.M., 2009. TERRA RNA binding to TRF2 facilitates heterochromatin formation and ORC recruitment at telomeres. Mol. Cell 35, 403–413.

Derecka, K., Balkwill, G.D., Garner, T.P., Hodgman, C., Flint, A.P., Searle, M.S., 2010. Occurrence of a quadruplex motif in a unique insert within exon C of the bovine estrogen receptor alpha gene (ESR1). Biochemistry 49, 7625–7633.

Flynn, R.L., Centore, R.C., O'Sullivan, R.J., Rai, R., Tse, A., Songyang, Z., et al., 2011. TERRA and hnRNPA1 orchestrate an RPA-to-POT1 switch on telomeric single-stranded DNA. Nature 471, 532–536.

Gomez, D., Guédin, A., Mergny, J.-L., Salles., B., Riou, J.-F., Teulade-Fichou, M.-P., et al., 2010. A G-quadruplex structure within the 5′-UTR of TRF2 mRNA represses translation in human cells. Nucleic Acids Res. 38, 7187–7198.

Huppert, J.L., Bugaut, A., Kumari, S., Balasubramanian, S., 2008. G-quadruplexes: the beginning and end of UTRs. Nucleic Acids Res. 36, 6260–6268.

Joachimi, A., Benz, A., Hartig, J.S., 2009. A comparison of DNA and RNA quadruplex structures and stabilities. Bioorg. Med. Chem. 17, 6811–6815.

Kumari, S., Bugaut, A., Balasubramanian, S., 2008. Position and stability are determining factors for translation repression by an RNA G-quadruplex-forming sequence within the 5′ UTR of the NRAS proto-oncogene. Biochemistry 47, 12664–12669.

Lacroix, L., Séosse, A., Mergny, J.-L., 2011. Fluorescence-based duplex-quadruplex competition test to screen for telomerase RNA quadruplex ligands. Nucleic Acids Res. 39, e21.

López de Silanes, I., d'Alcontres, M.S., Blasco, M.A., 2010. TERRA transcripts are bound by a complex array of RNA-binding proteins. Nat. Commun. 1, 1–9.

Luke, B., Lingner, J., 2009. TERRA: telomeric repeat-containing RNA. EMBO J. 28, 2503–2510.

Martadinata, H., Phan, A.T., 2009. Structure of propeller-type parallel-stranded RNA G-quadruplexes, formed by human telomeric RNA sequences in K^+ solution. J. Amer. Chem. Soc. 131, 2570–2578.

Mergny, J.L., De Cian, A., Ghelab, A., Saccà, B., Lacroix, L., 2005. Kinetics of tetramolecular quadruplexes. Nucleic Acids Res. 33, 81–94.

Morris, M.J., Basu, S., 2009. An unusually stable G-quadruplex within the 5′-UTR of the MT3 matrix metalloproteinase mRNA represses translation in eukaryotic cells. Biochemistry 48, 5313–5319.

Morris, M.J., Negishi, Y., Pazsint, C., Schonhoft, J.D., Basu, S., 2010. An RNA G-quadruplex is essential for cap-independent translation initiation in human VEGF IRES. J. Amer. Chem. Soc. 132, 17831–17839.

Nandakumar, J., Podell, E.R., Cech, T.R., 2010. How telomeric protein POT1 avoids RNA to achieve specificity for single-stranded DNA. Proc. Natl. Acad. Sci. USA 107, 651–656.

Parkinson, G.N., Lee, M.P.H., Neidle, S., 2002. Crystal structure of parallel quadruplexes from human telomeric DNA. Nature 417, 876–880.

Porro, A., Feuerhahn, S., Reichenbach, P., Lingner, J., 2010. Molecular dissection of telomeric repeat-containing RNA biogenesis unveils the presence of distinct and multiple regulatory pathways. Mol. Cell. Biol. 30, 4808–4817.

Randall, A., Griffith, J.D., 2009. Structure of long telomeric RNA transcripts: the G-rich RNA forms a compact repeating structure containing G-quartets. J. Biol. Chem. 284, 13980–13986.

Redon, S., Reichenbach, P., Lingner, J., 2010. The non-coding RNA TERRA is a natural ligand and direct inhibitor of human telomerase. Nucleic Acids Res. 38, 5797–5806.

Rzuczek, S.G., Pilch, D.S., Liu, A., Liu, L., LaVoie, E.J., Rice, J.E., 2010. Macrocyclic pyridyl polyoxazoles: selective RNA and DNA G-quadruplex ligands as antitumor agents. J. Med. Chem. 53, 3632–3644.

Satyanarayana, M., Kim, Y.A., Rzuczek, S.G., Pilch, D.S., Liu, A.A., Liu, L.F., et al., 2010. Macrocyclic hexa-oxazoles: influence of aminoalkyl substituents on RNA and DNA G-quadruplex stabilization and cytotoxicity. Bioorg. Med. Chem. Lett. 20, 3150–3154.

Schoeftner, S., Blasco, M.A., 2008. A "higher order" of telomere regulation: telomere heterochromatin and telomeric RNAs. EMBO J. 28, 2323–2336.

Schoeftner, S., Blasco, M.A., 2009. Developmentally regulated transcription of mammalian telomeres by DNA-dependent RNA polymerase II. Nat. Cell Biol. 10, 228–236.

Shahid, R., Bugaut, A., Balasubramanian, S., 2010. The BCL-2 5′ untranslated region contains an RNA G-quadruplex-forming motif that modulates protein expression. Biochemistry 49, 8300–8306.

Sparapani, S., Haider, S.M., Doriam, F., Gunaratnam, M., Neidle, S., 2010. Rational design of acridine-based ligands with selectivity for human telomeric quadruplexes. J. Amer. Chem. Soc. 132, 12263–12272.

Wieland, M., Hartig, J.S., 2007. RNA quadruplex-based modulation of gene expression. Chem. Biol. 14, 757–763.

Xu, Y., Kaminaga, K., Komiyama, M., 2008. G-quadruplex formation by human telomeric repeat-containing RNA in Na$^+$ solution. J. Amer. Chem. Soc. 130, 11179–11184.

Xu, Y., Suzuki, Y., Ito, K., Komiyama, M., 2010. Telomeric repeat-containing RNA structure in living cells. Proc. Natl. Acad. Sci. USA 107, 14579–14584.

Zhang, D.-H., Zhi, G.-Y., 2010. Structure monomorphism of RNA G-quadruplex that is independent of surrounding condition. J. Biotechnol. 150, 6–10.

Zhang, D.H., Fujimoto, T., Saxena, S., Yu, H.Q., Miyoshi, D., Sugimoto, N., 2010. Monomorphic RNA G-quadruplex and polymorphic DNA G-quadruplex structures responding to cellular environmental factors. Biochemistry 49, 4554–4563.

9

Design Principles for Quadruplex-binding Small Molecules

Overview

We discuss in this chapter some future directions for the design, optimization and evaluation of quadruplex-targeted small-molecules as drugs, and in particular how the features of quadruplexes may be best exploited for the design of biologically more effective quadruplex-binding ligands. The ultimate objective for much of the current intense activity in this area is to develop compounds suitable for clinical evaluation and eventual licensing for human use. With this in mind, a quadruplex drug discovery program needs to focus on:

1. Improving drug-like features.
2. Enhancing quadruplex affinity.
3. Enhancing quadruplex selectivity over other nucleic acid structures, especially over duplex DNA, which is the dominant form of DNA structure in a cell.
4. Enhancing selectivity for a particular quadruplex target compared to other quadruplexes.
5. Establishing *in vitro*, cell-based and *in vivo*, assay cascades for hit-to-lead selection.

Features Desired in Lead Ligand Scaffolds

The above aims are to some extent inter-dependent and good design will be focused on optimizing all of these features—there is little to be gained by enhancing quadruplex affinity and selectivity yet retaining those structural features that militate against eventual use as therapeutic agents. It is thus important for future molecular design studies in this area to have in hand at the outset appropriate lead scaffold motifs that possess drug-like features to at least some extent, and conversely do not have those characteristics that are well established to cause problems later on during the drug development process (see below). It is an axiom of drug development that attention to the pharmacokinetic issues of ADME (absorption, distribution, metabolism and excretion) is key to whether a drug candidate molecule can actually be an effective clinical agent.

Therapeutic Applications of Quadruplex Nucleic Acids. DOI: 10.1016/B978-0-12-375138-6.00009-1

Looking back at the large number of ligands developed to date (Chapters 5 and 6), one sees that the field is dominated by the constant theme of planar extended aromatic and heteroaromatic chromophores (from Sun *et al.*, 1997 onwards: see also Monchaud & Teulade-Fichou, 2008; Tan *et al.*, 2008). The reason for this domination is clear, since the simple concept of π–π stacking predicts that these motifs will interact effectively with a G-quartet. However, such chromophores also have a marked tendency to bind avidly and indiscriminately to a wide range of cellular macromolecules by virtue of their large hydrophobic surfaces, notably to serum albumin, which is among the most abundant of plasma proteins. These polycyclic compounds are often highly lipophilic and so may also accumulate in cell membranes and disrupt cell wall architecture. A further important reason for the desire to move away from extended polycyclic planarity is that this feature is found in classic duplex DNA intercalating compounds such as the cytotoxic anti-cancer drug Adriamycin. This has high levels of generalized toxicity and thus low tumor selectivity even though it is still a major clinically useful drug. Although duplex DNA binding for a given planar quadruplex-binding ligand can be minimized or is even undetectable in many *in vitro* experiments, there remains the nagging concern of some off-target binding within the context of the overwhelming excess of duplex DNA in a cell. It is encouraging for the future of this area that a number of synthetic non-polycyclic compounds are currently being developed, and that these are showing promise, at least in terms of selectivity. Their biological effects remain largely unexplored beyond simple assays of cell viability.

The Lipinski Rule of Five (Lipinski *et al.*, 2001), which is so often misused and misunderstood, was originally conceived to aid the development of orally bioavailable drugs, and was not designed to guide the medicinal chemistry development of all small-molecule drugs. Oral administration is a desirable objective for the treatment of a number of cancers, but is by no means an absolute requirement. So the tenets of the Lipinski Rule (molecular weight < 500, not more than five hydrogen bond donors, not more than ten hydrogen bond acceptors and a partition coefficient (log P) value < 5) have much to commend but do not need to be slavishly followed for many therapeutic projects. What is most important in the context of quadruplex-binding ligands is to pay especial attention to:

1. Minimization of ligand molecular weight. Aiming for compounds that approach the Lipinski limit of 500 daltons is clearly desirable since this can only improve uptake, although there are many examples of significantly larger molecules that are efficiently transported into cells.
2. Optimization of permeability and uptake. Small molecules targeting telomeric or promoter quadruplexes have to transverse both the outer and the nuclear membranes. Ideally log P and pK_a values should be obtained experimentally, but there are a large number of software packages that provide reliable estimates and that are very useful in the early stages of motif selection and then in lead optimization. Several programs are freely available to use on the web, at, for example, www.chemaxon.com and www.molinspiration.com. Table 9–1 gives log P values for several established quadruplex ligands (Figure 9–1), including those for which *in vivo* anti-cancer activity has been

Table 9–1   Selected physico-chemical properties for some representative G-quadruplex ligands, calculated using the facilities on the www.molinspiration.com website

Compound	Log P	Mol. wt (daltons)	Polar surface area (Å^2)	Volume (Å^3)	Reference
BRACO-19	1.86	597	96	575	Read *et al.*, 2001
AS1410	1.31	604	93	556	Martins *et al.*, 2007
Telomestatin	1.76	582	194	440	Kim *et al.*, 2002
Diaryl polyamide derivative	0.94	554	122	500	Rahman *et al.*, 2009
RHPS4	−0.04	333	9	287	Cookson *et al.*, 2005
6-Substituted indenoisoquinoline derivative	−0.85	361	42	340	Bejugam *et al.*, 2010
Bis-phenylethynyl-amide derivative	−0.99	596	44	585	Dash *et al.*, 2008
Quarfloxin	−0.28	605	97	536	Drygin *et al.*, 2009

reported. It is notable that none of these compounds fully complies with the Rule of Five. For example, the polycyclic acridine compound RHSP4 (Cookson *et al.*, 2005) has the advantage of a low molecular weight and is readily taken up into the nuclei of a wide range of cancer cells (see Chapter 6) in spite of its relatively poor log P score. Other large polycyclic ligands carrying several positive charges such as BRACO-19 (3$^+$) and the porphyrin TMPyP4 (4$^+$) are similarly readily taken up into cancer cells in culture and are also capable of diffusing into small experimental tumors to produce anti-cancer effects (see, for example, Burger *et al.*, 2005). It is plausible, though not proven, that their uptake is facilitated by an active transport mechanism in view of the high polarity of these compounds. However, their high charge hinders their ability to penetrate large tumor masses, so optimizing the lipophilic/lipophobic balance in a compound and monitoring log P values remains an important objective.

3. Avoidance of groupings that are readily metabolized, as well as those having inherent chemical instability. Ester groups can be cleaved by cellular esterases (unless this is part of a strategy to deliver a pro-drug or they are in sterically highly hindered positions). Ketones can be enzymatically reduced to hydroxyl groups. Amide groups can be cleaved by proteases. This is a potential concern since many G-quadruplex ligands have been developed over the past 12–14 years containing amide groups adjacent to phenyl rings (in order to extend the extended aromatic surface with minimal extra molecular weight cost). In many instances it appears that the close proximity to phenyl rings provides steric hindrance to proteolysis, but this cannot

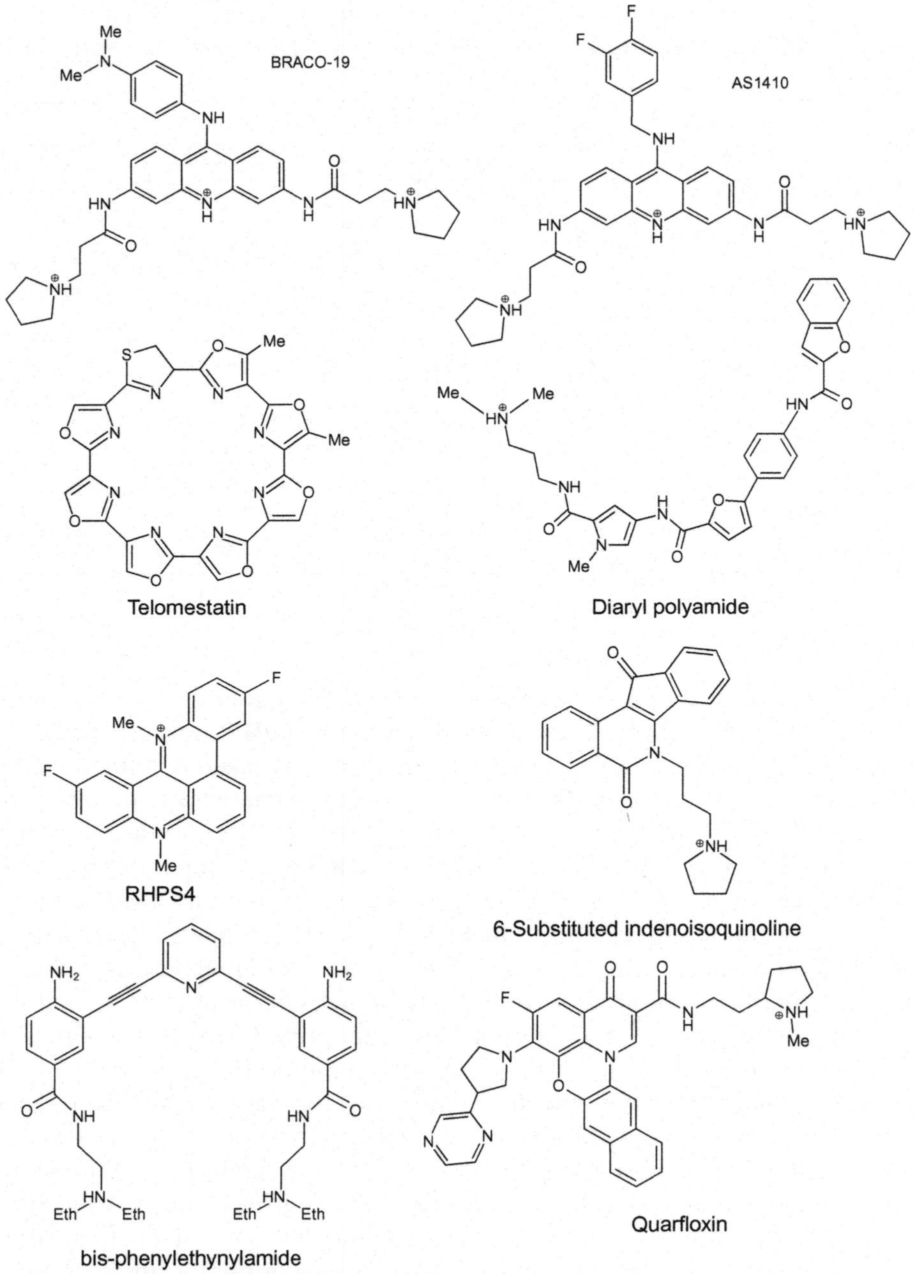

FIGURE 9–1 Structures of quadruplex-binding ligands highlighted in Table 9–1.

always be assumed. Methyl groups on aromatic systems can be oxidized to carboxylic acids. Aromatic nitro and amine groups can also be metabolized to toxic species, and should be avoided if possible. The presence of reactive alkylating groups such as activated epoxides can result in eventual cell death or even long-term carcinogenesis. The presence of phenolic or alcoholic hydroxyl groups can also be problematic since they can be conjugated to form glucuronides that are more polar and hence more readily cleared by the liver—this may sometimes be exploited to enhance the clearance of a highly non-polar compound.

4. Avoidance of groups that may be overtly toxic to a cell, for example certain metals such as nickel or lead, or thiol groups.

5. Off-target toxicity. This has not been reported, at least in publicly available documentation, for quadruplex ligands. It is a fair assumption that systematic screening in a range of neurological and cardiac-relevant assays has been undertaken for those few compounds that have undergone pre-clinical development, notably RHPS4, AS1410 and quarfloxin. Avoidance of even a low level of inhibition of the hERG potassium ion channel is especially important since hERG inhibition has been implicated in a drug-induced fatal cardiac arrhythmic syndrome and regulatory authorities now specifically request data from an hERG assay. An IC_{50} value for hERG of $>30\,\mu M$ is considered acceptable, indicating a likely low propensity for cardiac off-target effects.

The majority of drug candidates in cancer fail at later (and more costly) clinical trial stages because of a lack of statistically significant efficacy (Pearce *et al.*, 2008). So it is crucial not to compromise the physico-chemical factors at the expense of quadruplex affinity and biological activity at the target, as these are key determinants of likely therapeutic value. Chemical tractability at economic cost is also desirable for all therapeutic agents, as is ease and cost of scale-up for toxicology, clinical trials and eventual manufacture. Natural products have traditionally not been favored by the pharmaceutical industry because of the need for challenging and expensive multi-step syntheses and the resulting difficulties in generating analogs for structure–activity studies. However, the advent of a new generation of reagents and reactions, especially using transition metal catalysts, has transformed the potential of natural products as drug candidates since straightforward synthetic routes to them can increasingly be devised that contrast with the complex multi-person years of effort that have previously been the norm. This is exemplified by the recent total synthesis of the potent G-quadruplex ligand and telomerase inhibitor telomes-tatin, a natural product from *Streptomyces anulatus* (Linder *et al.*, 2011), together with several analogs, and suggests that further searches for potent lead quadruplex-binding compounds in Nature are warranted.

The Selection of Quadruplex-binding Core Motifs

The majority of quadruplex-binding ligands reported to date have been discovered by screening small focused compound libraries that at the outset have at least a superficial similarity to existing ligands of this class. Thus acridines were originally developed (Harrison *et al.*, 1999) as approximate structural mimetics of anthraquinones, which were the first group of compounds to be shown to have quadruplex affinity coupled with telomerase inhibitory activity (Sun *et al.*, 1997). The acridines also have the advantages of (i) improved aqueous solubility since the acridine central ring nitrogen is normally protonated at physiological pH, and (ii) enhanced quadruplex affinity since this cationic nitrogen atom can effectively form an extension of the central cation channel in a quadruplex, as subsequently visualized in several acridine-quadruplex crystal structures (Haider *et al.*, 2003: Campbell *et al.*, 2008, 2009, 2011).

Random screening of large chemical libraries has been very rarely used up till now in the search for new quadruplex-binding ligands, in striking contrast with their extensive use in other therapeutic areas. This is unsurprising when targeting telomeric quadruplexes since the essential component for this, a high-throughput functional assay for telomerase activity, is not generally available. On the other hand, many high-throughput assays for transcriptional or translational activity have been developed (see a recent example for screening libraries of potential inhibitors of WNT transcription: Ewan *et al.*, 2010), which are relevant to promoter and RNA quadruplexes, respectively.

Structural information from X-ray crystallography or NMR has been only rarely used for the rational design of improved quadruplex-binding ligands, in large part because of the lack of suitable structural starting points—only a very small number of quadruplex–ligand 3-D structures have been reported to date (Table 9–2). The majority of structure-based design studies have used native quadruplex structures as starting points. These may be sufficiently close to the structures of their ligand complexes for them to be useful in initial screening, but comparison of native and ligand-bound complexes shows that loop conformations in particular are readily changed on ligand binding (Figure 9–2). Thus unless the design study is on ligands that are very closely related in structural terms to that in an experimentally determined quadruplex–ligand complex, there is the possibility of significant changes to binding-site geometry that cannot readily be modeled, except perhaps by long time-scale or coarse-grained molecular dynamics simulations. In a recent example of conservative molecular design based on relevant crystal structure information, a structural mimetic of BRACO-19 was designed on the basis of retaining the concept of three arms of this trisubstituted acridine binding in three grooves of a quadruplex, as found in the BRACO-19-telomeric quadruplex crystal structure (Campbell *et al.*, 2008). The acridine core was replaced by a symmetrically trisubstituted phenyl ring (Figure 9–3a), with side-chains of appropriate length and bulk to fill each of the three grooves (Figure 9–3b). The optimal compounds were found experimentally to have greater affinity for the human telomeric quadruplex coupled with much lower affinity for duplex DNA compared to BRACO-19 itself (Lombardo *et al.*, 2010).

A number of studies have used *in silico* and computational approaches to analyze the behavior of existing ligands—these have been outlined in earlier chapters. Rather fewer studies have used *in silico* screening of compound libraries to help select lead compounds. Most of these have used native structures taken from the Protein Data Bank.

Table 9–2 Quadruplex–Ligand Structures available in the PDB (www.pdb.org). Individual ligands have been discussed in detail in Chapter 4. h-tel signifies a human telomeric nucleic acid sequence. *Oxy*-tel is a telomeric sequence from *Oxytricha nova* (the isomorphous crystal structures of a number of closely related disubstituted acridine derivatives complexed to the same sequence are listed in Chapter 4)

Sequence	Origin	Ligand	Method	PDB id	Reference
$(T_2AG_3T)_4$	h-tel	RHPS4	NMR	1NMZ	Gavathiotis *et al.*, 2003
$(G_4T_4G_4)_2$	*Oxy*-tel	Disubstituted acridine deriv	X-ray	1L1H	Haider *et al.*, 2003
$(TAG_3T_2AG_3T)_2$	h-tel	BRACO-19	X-ray	3CE5	Campbell *et al.*, 2008
$(TAG_3T_2AG_3T)_2$	h-tel	Naphthalene diimide deriv	X-ray	3CCO	Parkinson *et al.*, 2008
$(TAG_3T_2AG_3)_2$	h-tel	TMPyP4	X-ray	2HRI	Parkinson *et al.*, 2007
$TAG_3\ T_2AG_3T_2AG_3\ T_2AG_3$	h-tel	Naphthalene diimide derive	X-ray	3CDM	Parkinson *et al.*, 2008
$TGAG_3TG_2IGAG_3TG_4A_2G_2A$	*c-myc*	TMPyP4	NMR	2A5R	Phan *et al.*, 2005
$r(UAG_3U_2AG_3U)_2$	h-tel	Bis-triazole acridine deriv	X-ray	3MIJ	Collie *et al.*, 2011
$(TAG_3T_2AG_3T)_2$	h-tel	Ni- and Cu-salphen derivs	X-ray	3QSC 3QSF	Campbell *et al.*, to be published

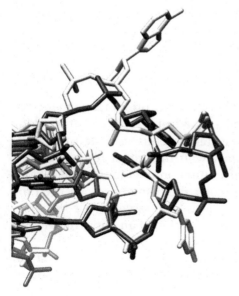

FIGURE 9–2 Superposition of one of the TTA loops in the native human telomeric quadruplex crystal structure (Parkinson *et al.*, 2002), colored black (PDB id 1KF1) with the same loop in the crystal structure of a naphthalene diimide complex (Parkinson *et al.*, 2008) (PDB id 1CDM), colored gray. Significant differences in the orientation of the adenine and first thymine bases are apparent. This and other figures have been drawn using the CHIMERA program (http://www.cgl.ucsf.edu/chimera) (Petterson *et al.*, 2004).

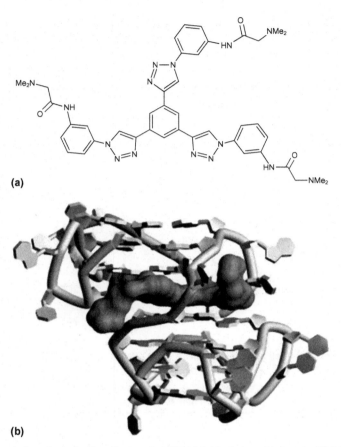

FIGURE 9–3 (a) A triazole-based trisubstituted mimetic of BRACO-19, designed using the BRACO-19 quadruplex co-crystal structure (Lombardo *et al.*, 2010). **(b)** A molecular model for the quadruplex–mimetic complex, showing the two quadruplexes constituting the biological unit in cartoon form, with the solvent-accessible surface of this mimetic molecule colored gray.

Two recent examples have targeted a quadruplex in the NHE III promoter sequence of the *c-myc* oncogene. The 3-D structure of this particular quadruplex was constructed for this study using the human telomeric quadruplex crystal structure as a starting point. Screening a 20,000 compound natural product library with this target resulted in the identification (Lee *et al.*, 2010) of the naphthopyrone natural product fonsecin B (Figure 9–4a), originally isolated from a fungal source. Molecular modeling has been used for lead optimization in the design of platinum (II) complexes, also targeting a *c-myc* promoter quadruplex (Wang *et al.*, 2010), following the identification of an initial hit with an *in vitro* polymerase stop assay screen.

A drug-like *in silico* library of over 100,000 compounds has been screened against the human telomeric quadruplex crystal structure itself (Ma *et al.*, 2008). The best compound

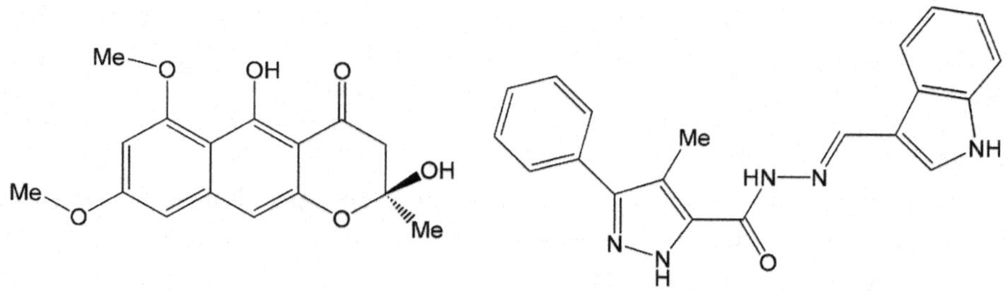

(a)   **(b)**

FIGURE 9–4 Two quadruplex-binding compounds found by *in silico* screening. **(a)** The natural product fonsecin B (Lee *et al.*, 2010). **(b)** A synthetic substituted indole (Ma *et al.*, 2008).

from this screen, a substituted indole (Figure 9–4b), has drug-like features with a molecular weight of 343 and a calculated log P of 3.9 (see below), although its aqueous solubility is likely to be low. Its quadruplex stabilizing ability, though not exceptional, is comparable with many other established ligands, as indicated by a ΔT_m value of 13.5° at a 1 μM concentration, and its affinity for duplex DNA appears to be low. A rather small commercially available and structurally very diverse library of 6000 compounds has been screened (Cosconati *et al.*, 2009) against the tetramolecular d(TGGGGT) quadruplex using the AUTODOCK program. The top 30 hits were then screened using changes in NMR chemical shift to locate the best six compounds, with the aim of finding groove-binding leads. The best compound appears to span the grooves and produce large changes in the NMR spectrum of the quadruplex, indicative of high affinity.

These promising results are supportive of *in silico* screening as a viable approach to finding novel new lead compounds for quadruplex ligand discovery. The very large publically available ZINC database (Irwin & Shoichet, 2005) (www.zinc.docking.org) currently has 36 million compounds in it, of which over 13 million can be purchased and therefore evaluated. There are as yet no reported uses of ZINC for finding new quadruplex-binding ligands.

Fragment-based approaches use libraries of very small, relatively low-affinity compounds (typically having molecular weights < 250–300 daltons, good aqueous solubility and < 3 hydrogen bond donors/acceptors), and usually contain just a few hundred carefully chosen compounds (Coyne, Scott & Abell, 2010; Feyfant *et al.*, 2011). These can be screened by high-throughput crystallography, NMR or *in silico* methods to locate where a fragment is bound. The fragment is then extended computationally to generate larger lead molecules that fill out the binding site and have in-built selectivity for it as well as good drug-like features from the outset. These can then be chemically synthesized and evaluated. The method has been especially successful in a number of protein crystallography structure-based drug design projects on, for example, cyclin-dependent kinase inhibitors (Tisi *et al.*, 2008), and is applicable in principle to nucleic acids with complex tertiary structures. Studies on ribosomal RNAs (Bodoor *et al.*, 2009) and riboswitches (Chen *et al.*, 2010) as targets have shown that it is possible in practice to generate high-quality fragment libraries for nucleic acids as well as proteins. Fragment-based drug design does have some practical challenges such as a requirement for a large number of crystals for a crystallographic approach, but has the advantage of rapidly generating highly specific leads with entirely novel chemical platforms that are not related to the existing ligand families.

Two principal categories of non-covalent binding site are generic to all quadruplex nucleic acids: (i) a terminal G-quartet suitable for π–π stacking, and (ii) a groove or loop fold for non-bonded/electrostatic interactions. The overwhelming majority of quadruplex ligands reported to date interact predominantly by π–π stacking, albeit with substituents sometimes located in a groove. Very few compounds have been authenticated as solely quadruplex groove binders. Even so, the groove mode has several attractions for quadruplex drug design:

1. Groove-binding molecules have been well studied as a DNA duplex binding motif (Neidle, 2008) and a large number of compounds and diverse structural platforms are available, with one compound reaching a Phase III clinical trial for trypanosomiasis (Paine *et al.*, 2010) and another in trials for human cancer (Alley *et al.*, 2004).

2. They tend to have more drug-like features, since they do not contain extended polycyclic aromatic hydrocarbon-like frameworks. One important caveat is that the majority of "simple" duplex DNA groove-binding compounds are AT-selective, whereas the quadruplex grooves have inherently G-rich features at the floor of the grooves with the guanine N2 substituent accessible for providing hydrogen bond donation capability (Figure 9–5). Suitable motifs may be based on, for example, the polyamide motif, which can be in principle tailored for a wide range of G-containing sequences (Hsu *et al.*, 2007). Another potential ligand class is suggested by the findings that aminoglycoside antibiotics such as neomycin and paromomycin bind with moderately high affinity (K_a ~10^5 mol^{-1}) to *Oxytricha nova* telomeric quadruplexes (Ranjan *et al.*, 2010), and *in silico* docking studies suggest that they bind in the wide groove of this quadruplex type.

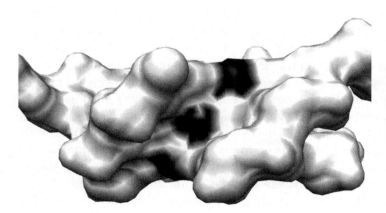

FIGURE 9–5 View of the solvent-accessible surface of a quadruplex groove. The surface of the guanine N2 substituents is colored black, highlighting their accessibility to ligand binding.

B-DNA groove-binding features in a molecule emphasize either linearity or a curved, concave surface. They can be redesigned to result in a preference for G-quartet binding, as has been shown, for example, with a small library of diaryl polyamide compounds (Rahman *et al.*, 2009), in which the particular pattern of amide linkages on the individual rings (Figure 9–1) ensures that the molecule has two sharp turns and thus an overall U-shape. Molecular modeling has predicted that this shape gives good overlap with three out of the four guanines in a G-quartet (Figure 9–6), yet poor complementarity to the convex minor groove surface of duplex B-form DNA. The overlap arrangement (Figure 9–6) has similarities to that predicted (Linder *et al.*, 2011) for the cyclic natural product telomestatin (Figure 9–7), which has high affinity and selectivity for several quadruplex nucleic acids (Kim *et al.*, 2002; Nielsen & Ulven 2010; Linder *et al.*, 2011). Both telomestatin and several diaryl polyamides compounds have low affinity for duplex DNA since their shapes are incompatible with either B-DNA groove or intercalative binding.

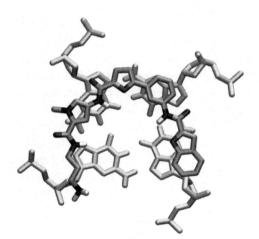

FIGURE 9–6 A stick representation from molecular modeling of the interaction between a diaryl polyamide derivative (colored gray) and a terminal G-quartet (adapted from Rahman *et al.*, 2009).

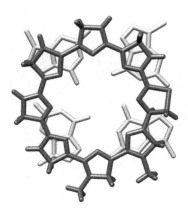

FIGURE 9–7 A stick representation from molecular modeling of the interaction between a telomestatin molecule (colored gray) and a terminal G-quartet.

The favorable quadruplex-binding profile and promising biological activity of telomestatin have prompted a search for other synthetic macrocyclic platforms (Granzhan *et al.*, 2010; Monchaud *et al.*, 2010; Nielsen & Ulven, 2010), although these remain to be explored for drug-like activity and features.

There are some opportunities for selectivity even at the G-quartet level since the available surface area does differ with distinct overall topologies, and is especially dependent on whether lateral or diagonal loops cover a quartet surface. The available surface area is maximum for propeller-looped, i.e. all-parallel, structures (Figure 9–8) and, for example, the human parallel quadruplex has good accessibility for both 3′ and 5′ terminal G-quartets in the human parallel quadruplex whereas only the G-quartet at one end is fully accessible in the (3 + 1) anti-parallel structure (Dai *et al.*, 2006).

Groove widths in several experimentally determined quadruplex structures are compared in Table 9–3, which highlights the high variability between them and is a consequence of a number of factors: glycosidic conformation, the loop type and length, as well as sequence. Parallel loops normally require all guanosine glycosidic angles to be in an *anti* conformation, whereas anti-parallel loops have mixed *syn* and *anti* glycosidic conformations. These differences are reflected in the high symmetry of the all-parallel native human telomeric structure compared to these others. Differences in grooves between the structures are most apparent for promoter quadruplexes, which have wide diversity in primary sequence and in tertiary structure, as exemplified by the *c-myc*, *bcl*-2 and *c-kit*1 quadruplexes (Figure 9–9). This diversity in groove widths is in striking contrast with the narrow minor groove in duplex B-DNA (Figure 9–10), which has limited sequence dependence, and an average width of 6Å. There is some evidence (Todd *et al.*, 2007) that the existence of a stable quadruplex fold for a sequence that has a unique occurrence in the human genome (for example, for *c-kit*1) implies structural uniqueness as well, and thus enhanced target druggability. The large differences in groove widths between

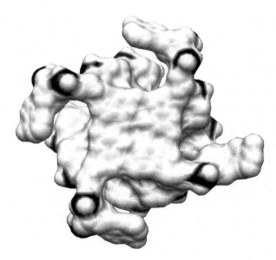

FIGURE 9–8 Solvent-accessible surface representation of a terminal G-quartet in the human telomeric quadruplex crystal structure, highlighting its accessibility for ligand binding. The surfaces of the phosphorus atoms have been colored black.

Table 9–3 Groove widths in various intramolecular quadruplex structures, in Å, defined in terms of minimum inter-strand phosphate oxygen...oxygen distances

Structure	PDB id	Groove 1	Groove 2	Groove 3	Groove 4	Reference
Human telomeric parallel (K⁺)	1KF1	10.0	11.4	11.8	11.4	Parkinson *et al.*, 2002
*Human telomeric (3+1)	2HY9	3.4	12.1	10.8	9.2	Dai *et al.*, 2006
*Human telomeric (Na⁺)	143D	8.7	9.8	13.9	18.4	Wang & Patel, 1993
*c-kit1	2O3M	6.1	11.1	5.8	14.6	Phan *et al.*, 2007
*c-myc	1XAV	9.7	14.9	12.7	11.9	Ambrus *et al.*, 2005
*bcl-2	2F8U	10.3	6.4	17.5	7.7	Dai *et al.*, 2006
Human telomeric parallel + ND ligand	3CDM	7.8	8.4	8.5	11.6	Parkinson *et al.*, 2008

*are from NMR structure determinations and calculations have been made on individual structures taken from the ensemble of structures deposited in the PDB (www.pdb.org).

quadruplexes highlighted in Table 9–3 also suggest that these features are exploitable for the future design of selective ligands, for *in silico* library screening in search of quadruplex selectivity, or for modifying existing compounds. The crystal structure of a ND (naphthalene diimide) ligand complexed with an all-parallel human telomeric quadruplex shows that the ligand binding has induced some changes in groove dimensions compared to the native structure (Parkinson *et al.*, 2002, 2008), indicating that ligand binding can perturb other aspects of quadruplex structure as well as loop conformations (Figure 9–2). The narrowing of a groove in this structure adjacent to one of the substituents (Figure 9–11) has been exploited to design an extended side-chain terminating in an N-methyl piperazine ring that is bulkier than the original N-dimethyl end-group (Parkinson *et al.*, 2008), and would better fit into this groove (Hampel *et al.*, 2010), thus enhancing quadruplex affinity and selectivity (Figure 9–12). Experimental studies on this analog have borne out these predictions (Hampel *et al.*, 2010).

Quadruplex loops have not been explicitly targeted by small molecule ligands to date. However, they can form significant structural elements, especially when they comprise more than three nucleotides, and are the principal components of quadruplex-specific cavities that are in addition to the four grooves generic to all normal quadruplexes. The structure of the *c-kit1* quadruplex both in solution (Phan *et al.*, 2007) and in the crystalline state (Wei *et al.*, 2011, to be published) shows that its unique topology includes a pronounced five-nucleotide AGGAG stem-loop, which is almost perpendicular to its adjacent terminal G-quartet and together forms a pronounced structural cleft of a size appropriate for small-molecule binding (Figure 9–13).

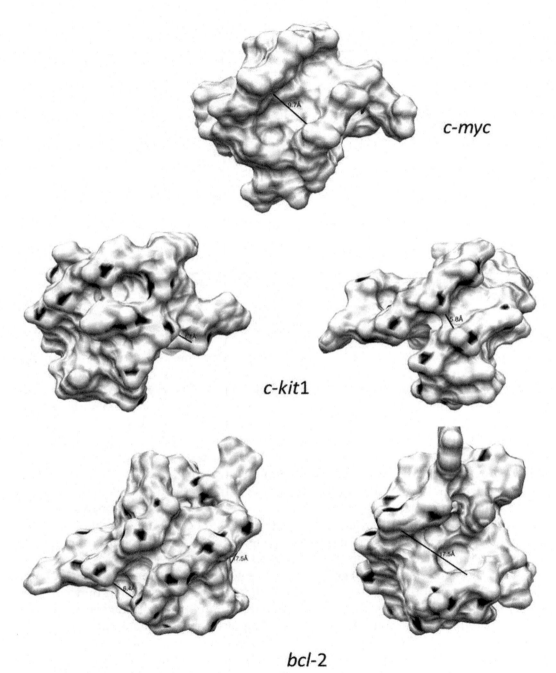

FIGURE 9–9 Solvent-accessible surface views of grooves in selected quadruplex structures, taken from the PDB coordinates.

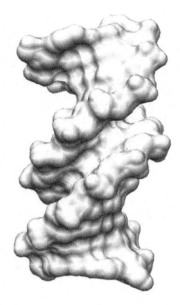

FIGURE 9–10 Solvent-accessible surface view into the minor groove of B-DNA.

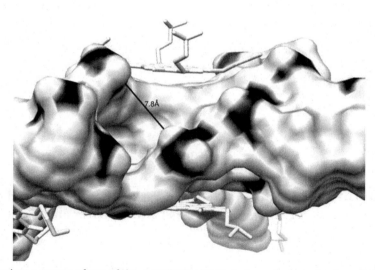

FIGURE 9–11 View into a groove of one of the crystallographically independent quadruplexes in the naphthalene diimide quadruplex complex structure (Parkinson, Ghosh & Neidle, 2007).

Quadruplex selectivity over duplex DNA (or RNA) is inherent in the differences in numbers of grooves in their structures. Duplex nucleic acids have two grooves, with B-DNA, the major physiological form, having a narrow yet deep minor groove and a wide, shallow major groove. The four grooves in a quadruplex thus provide an obvious basis for selectivity, and molecules can readily be designed with three or more substituents

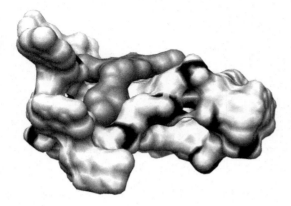

FIGURE 9–12 Computer modeling representation of the groove interactions of an N-methyl-piperazine group attached to a side-chain of a naphthalene diimide ligand, using the ligand crystal structure as a template (Parkinson, Cuenca & Neidle, 2008).

AGGAG loop

cleft

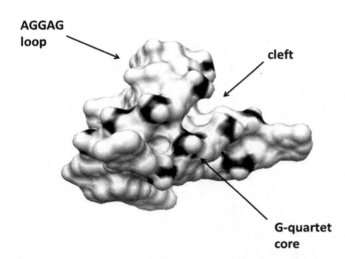

FIGURE 9–13 Solvent-accessible surface representation of the *c-kit*1 structure, with phosphorus atoms colored black, viewed along the G-quartet planes. The cleft between the stem-loop and the adjacent G-quartet is ~4Å wide at its narrowest point.

G-quartet core

to do this. BRACO-19, for example, has three side-arms that reach out and contact three grooves (Read *et al.*, 2001: Campbell *et al.*, 2008). Four grooves are accessed simultaneously by a manganese porphyrin derivative (Figure 9–14) with extended cationic side-chains (Dixon *et al.*, 2007), which has over 10^3-fold selectivity over duplex DNA whereas the parent porphyrin TMPyP4 with much shorter arms has only limited selectivity.

The Selection of Ligand Substituents

A relatively narrow set of side-chain substituents has been studied across the range of polycyclic ligands. In very general terms acyclic side-chains with two or three carbon atoms and terminating in a cationic charged group such as NMe_2, $NEth_2$, pyrrolinide, piperidine or piperazine have found favor. Carbon-chain length requirements depend on the size of the polycyclic platform to which they are attached. These charged groups also

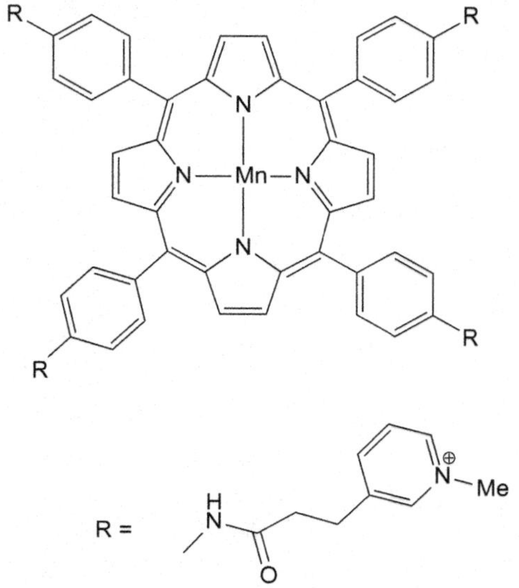

FIGURE 9–14 Structure of a substituted manganese porphyrin (Dixon *et al.*, 2007).

improve aqueous solubility, and are therefore especially useful when attached to many otherwise insoluble polycyclic heteroaromatic motifs. All of these groups are protonated at physiological pH, and the dogma has arisen that quadruplex-binding structure–activity relationships are optimal with these substituents: conversely, morpholino groups, even though they can readily fit into a quadruplex groove, tend to destabilize or minimize binding. Ligands with between one and four protonated side-chains have been reported, and quadruplex affinity for many polycyclic derivatives tends to increase with an increasing number of such side-chains, albeit frequently at the expense of selectivity. It is widely assumed that these protonated groups complement anionic phosphate charges, although it is rare to observe direct cationic group-phosphate contacts in the crystal structures. Increasing the number of cationic charges and hence the polarity of a ligand does not improve drug-like characteristics, and the major pharmacokinetics parameter, *in vivo* half-life and elimination rate, can become unacceptably short. To counter this it is possible to more subtly engineer and modify charged groups by systematically adding hydrophobic atoms and so increase log P values in a logical manner (Figure 9–15).

Substituents on a particular quadruplex binding platform are not always required to produce high quadruplex affinity, as has been demonstrated by the favorable properties of telomestatin and other macrocyclic ligands. However, aqueous solubility is a problem for telomestatin and judicious addition of small side-chains is warranted.

Targets and Tools for Ligand Selectivity and Discovery

In principle it is desirable to target a ligand to a given quadruplex with high affinity yet with minimal affinity for either duplex DNA or other quadruplexes. There are now a considerable number of ligands whose duplex affinity is at least several orders of magnitude

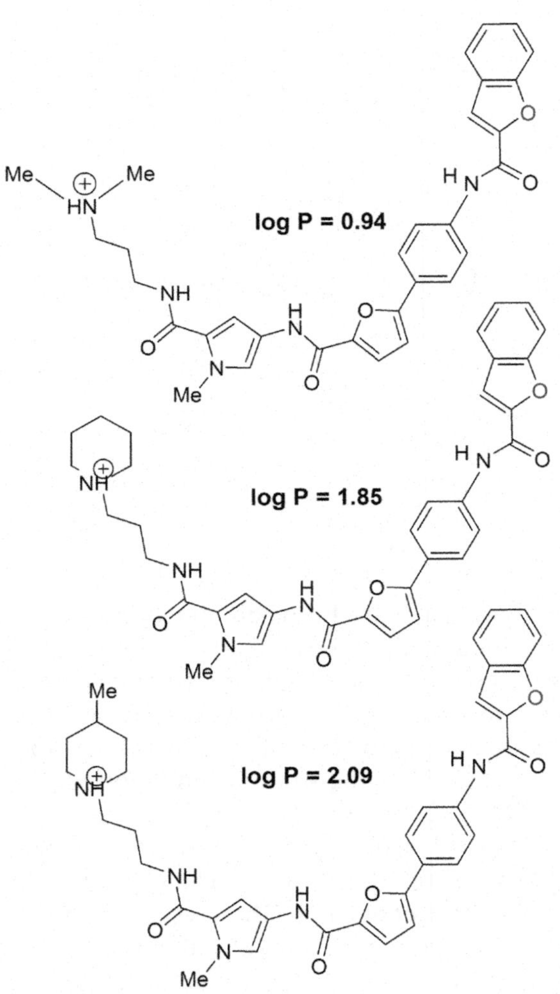

FIGURE 9–15 Effects of systematically changing the cationic end-group in a diaryl polyamide ligand. Log P values have been calculated using the facilities on the www.molinspiration.com website.

less than that to a quadruplex. Since there is still very little known about the overwhelming majority of genomic DNA and RNA quadruplexes, or even whether a given sequence forms a stable structure, the question arises as to whether progress can yet be made. In the case of telomeric quadruplexes selectivity arises because the target may not be a single quadruplex, but rather several formed along the single-stranded DNA overhang (see, for example, Haider & Neidle, 2009 for details of a parallel quadruplex-based tandem repeat model). So the ligand binding sites are internal quasi-intercalative ones (Figure 9–16) rather than the more open sites likely to be present in genomic quadruplexes. The overwhelming majority of telomeric quadruplex binding studies, however, still use a single intramolecular quadruplex comprising four telomeric DNA (or RNA) repeats so this model may not be fully representative of quadruplex formation in a telomere, and quadruplex dimers (as in

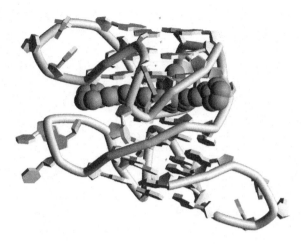

FIGURE 9–16 Molecular model of the BRACO-19 ligand bound to a two-quadruplex model comprising eight TTA repeats, constructed (Haider & Neidle, 2009) from the BRACO-19–quadruplex crystal structure (Campbell *et al.*, 2008). The DNA is shown in cartoon form and the BRACO-19 molecule is shown in CPK representation, colored gray.

Figure 9–16) may represent the smallest relevant biological unit, in accord with a ligand-binding study using an eight human telomere-repeat sequence (Bai *et al.*, 2008).

In the absence of information on most putative quadruplexes, an appropriate strategy would be first to decide to focus on a particular disease or pathway, then on a key driver such as a dominant oncogene, which is confirmed by a bioinformatics analysis to possess at least one putative quadruplex motif in a promoter or untranslated region. Selection between these motifs, or motifs from a small number of oncogenes, should then be on the basis of potential sequence uniqueness and a demonstration of stable quadruplex formation *in vitro*. Key indicators of this are available from characteristic behavior in the following:

1. Melting experiments (Mergny & Lacroix, 2009) at 295 nm, showing a decrease in absorbance on heating, with a single transition (the ΔT_m) being indicative of a single quadruplex species.
2. Circular dichroism (Paramasivan, Rujan & Bolton, 2007), which can be useful in assigning overall topology. A positive CD band around 260 nm and a minimum at around 240 nm indicates a parallel quadruplex topology whereas a positive band around 290–295 nm with a negative band near 260 nm indicates an anti-parallel arrangement.
3. A 1-D NMR experiment and the observation of imino proton signals (Webba da Silva, 2007). Guanine imino resonances at 10.0–12.0 ppm are characteristic of a G-quartet hydrogen-bonding arrangement. Formation of a single quadruplex species can be inferred from the number and intensity of the peaks in this region.

Conclusions

This chapter has outlined the ways in which ligand discovery can be approached, emphasizing the structure-based concepts that have been described throughout the

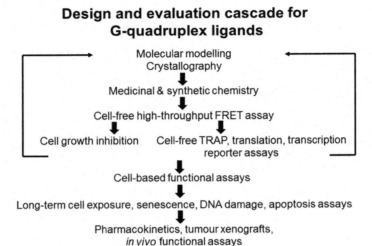

FIGURE 9–17 A drug discovery evaluation cascade for quadruplex-binding ligands. Many of the steps are generic whether the target is a promoter, an RNA or a telomeric quadruplex. It may even be worthwhile at the outset using a catch-all approach, and subsequently to examine selectivity. The functional assays will vary, depending on the biological context of the target; the reporter assays, such as have been developed for a range of promoter and RNA quadruplexes (see, for example, Beaudoin & Perreault, 2010) enable quadruplex binding and effects (for example, on transcription or translation) to be assessed in a high-throughput manner. The *in vitro* binding FRET assay is also high throughput (Mergny *et al.*, 2001; De Cian *et al.*, 2005), but the TRAP telomerase assay is not. A more recent and potentially very useful development is a fluorescence-based competition assay for evaluating quadruplex-duplex selectivity (Lacroix *et al.*, 2010). High-throughput non-PCR-based telomerase assays are currently being developed (for example, see Cristofari *et al.*, 2007) but as yet are not widely used.

book. Figure 9–17 summarizes the flow of assays for these early steps in quadruplex-targeted anti-cancer drug lead discovery and evaluation. These are just one part of what is needed to develop a new therapeutic agent, and many of the succeeding steps are generic to drug discovery. The important choice of biomarkers in clinical trials will need to reflect the biology of the particular quadruplex being targeted, for example changes in *c-myc* or *c-kit* expression for these oncogenes, or DNA damage responses/telomerase activity for telomeric targeting, when developing new anti-cancer agents.

Most pre-clinical drug discovery in the quadruplex field has to date been undertaken in academic laboratories, notable exceptions being the development of the anti-cancer drug quarfloxin by the Hurley laboratory and Cylene Pharmaceuticals (www.cylenepharma .com), the earlier work on several quinoline-based and related ligands by the French company Aventis and their academic collaborators (see, for example, Riou *et al.*, 2002), the development of trisubstituted acridines based on BRACO-19 (Martins *et al.*, 2007) by my own laboratory in collaboration with Antisoma plc (www.antisoma.com), and the continuing development of telomere targeting agents by Pharminox plc (www.pharminox .com) based on the RHPS4 molecule devised by the Stevens laboratory (Cookson, Heald & Stevens, 2005). The sole quadruplex-binding small molecule that has entered clinical evaluation to date is quarfloxin, developed by Cylene Pharmaceuticals. Such undertakings are currently beyond the capabilities of academic laboratories, although as the focus of drug discovery is increasingly moving from industry to academia, it is to be hoped that further effective university–industry consortia will be established to bring new compounds into the clinic, not only for human cancers but also for other major diseases. We are now at the point where the G-quadruplex concept is becoming accepted by the wider biological community, and it is an appropriate time to consider where it is heading. The field has concentrated to date almost exclusively on quadruplex ligands as anti-cancer agents, but bioinformatics tell us that there are many non-oncogenic genomic quadruplex targets that have relevance in other types of human diseases. The possibilities of exploiting quadruplexes in other eukaryotic and prokaryotic genomes such as in parasitic or infectious agents have not yet been seriously explored; a pioneering study on quadruplexes in the malaria genome (De Cian *et al.*, 2008) indicates the potential as well as the challenges of overcoming problems of selectivity.

References

Alley, M.C., Hollingshead, M.G., Pacula-Cox, C.M., Waud, W.R., Hartley, J.A., Howard, P.W., et al., 2004. SJG-136 (NSC 694501), a novel rationally designed DNA minor groove interstrand cross-linking agent with potent and broad spectrum antitumor activity: part 2: efficacy evaluations. Cancer Res. 64, 6700–6706.

Bai, L.-P., Hagihara, M., Jiang, Z.-H., Nakatani, K., 2008. Ligand binding to tandem G quadruplexes from human telomeric DNA. Chem. Bio. Chem. 9, 2583–2587.

Beaudoin, J.D., Perreault, J.P., 2010. 5'-UTR G-quadruplex structures acting as translational repressors. Nucleic. Acids Res. 38, 7022–7036.

Bejugam, M., Gunaratnam, M., Müller, S., Sanders, D.A., Sewitz, S., Fletcher, J., et al., 2010. Targeting the c-Kit promoter quadruplexes with 6-substituted indenoisoquinolines, a novel class of G-quadruplex stabilising small molecule ligands. ACS Med. Chem. Lett. 1, 306–310.

Bodoor, K., Boyapati, V., Gopu, V., Boisdore, M., Allam, K., Miller, J., et al., 2009. Design and implementation of a ribonucleic acid (RNA) directed fragment library. J. Med. Chem. 52, 3753–3761.

Burger, A.M., Dai, F., Schultes, C.M., Reszka, A.P., Moore, M.J., Double, J.A., et al., 2005. The G-quadruplex-interactive molecule BRACO-19 inhibits tumor growth, consistent with telomere targeting and interference with telomerase function. Cancer Res. 65, 1489–1496.

Campbell, N.H., Parkinson, G.N., Reszka, A.P., Neidle, S., 2008. Structural basis of DNA quadruplex recognition by an acridine drug. J. Amer. Chem. Soc. 130, 6722–6724.

Campbell, N.H., Patel, M., Tofa, A.B., Ghosh, R., Parkinson, G.N., Neidle., S., 2009. Selectivity in ligand recognition of G-quadruplex loops. Biochemistry 48, 1675–1680.

Campbell, N.H., Smith, D.L., Reszka, A.P., Neidle, S., O'Hagan, D., 2011. Fluorine in medicinal chemistry: β-fluorination of peripheral pyrrolidines attached to acridine ligands affects their interactions with G-quadruplex DNA. Org. Biomol. Chem. 9, 1328–1331.

Chen, L., Cressina, E., Leeper, F.J., Smith, A.G., Abell, C., 2010. A fragment-based approach to identifying ligands for riboswitches. ACS Chem. Biol. 5, 355–358.

Collie, G.W., Sparapani, S., Parkinson, G.N., Neidle, S., 2011. Structural basis of telomeric RNA quadruplex-acridine ligand recognition. J. Amer. Chem. Soc. 133, 2721–2728.

Cookson, J.C., Heald, R.A., Stevens, M.F., 2005. Antitumor polycyclic acridines. 17. Synthesis and pharmaceutical profiles of pentacyclic acridinium salts designed to destabilize telomeric integrity. J. Med. Chem. 48, 7198–7207.

Cosconati, S., Marinelli, L., Trotta, R., Virno, A., Mayol, L., Novellino, E., et al., 2009. Tandem application of virtual screening and NMR experiments in the discovery of brand new DNA quadruplex groove binders. J. Amer. Chem. Soc. 131, 16336–16337.

Coyne, A.G., Scott, D.E., Abell, C., 2010. Drugging challenging targets using fragment-based approaches. Curr. Opin. Chem. Biol. 14, 299–307.

Cristofari, G., Reichenbach, P., Regamey, P.O., Banfi, D., Chambon, M., Turcatti, G., et al., 2007. Low- to high-throughput analysis of telomerase modulators with Telospot. Nature Methods 4, 851–853.

Dai, J., Punchihewa, C., Ambrus, A., Chen, D., Jones, R.A., Yang, D., 2006. Structure of the intramolecular human telomeric G-quadruplex in potassium solution: a novel adenine triple formation. Nucleic Acids Res. 35, 2440–2450.

Dash, J., Shirude, P.S., Hsu, S.-T.D., Balasubramanian, S., 2008. Diarylethynyl amides that recognize the parallel conformation of genomic promoter DNA G-quadruplexes. J. Amer. Chem. Soc. 130, 15950–15956.

De Cian, A., Grellier, P., Mouray, E., Depoix, D., Bertrand, H., Monchaud, D., et al., 2008. Plasmodium telomeric sequences: structure, stability and quadruplex targeting by small compounds. Chem. Bio. Chem 9, 2730–2739.

De Cian, A., Guittat, L., Shin-Ya, K., Riou, J.-F., Mergny, J.-L., 2005. Affinity and selectivity of G4 ligands measured by FRET. Nucleic Acids Symp. Ser., 235–236.

Dixon, I.M., Lopez, F., Tejera, A.M., Estève, J.P., Blasco, M.A., Pratviel, G., et al., 2007. A G-quadruplex ligand with 10 000-fold selectivity over duplex DNA. J. Amer. Chem. Soc. 129, 1502–1503.

Drygin, D., Siddiqui-Jain, A., O'Brien, S., Schwaebe, M., Lin, A., Bliesath, J., et al., 2009. Anticancer activity of CX-3543: a direct inhibitor of rRNA biogenesis. Cancer Res. 69, 7653–7661.

Ewan, K., Pajak, B., Stubbs, M., Todd, H., Barbeau, O., Quevedo, C., et al., 2010. A useful approach to identify novel small-molecule inhibitors of Wnt-dependent transcription. Cancer Res. 70, 5963–5973.

Feyfant, E., Cross, J.B., Paris, K., Tsao, D.H., 2011. Fragment-based drug design. Methods Mol. Biol. 685, 241–252.

Gavathiotis, E., Heald, R.A., Stevens, M.F., Searle, M.S., 2003. Drug recognition and stabilisation of the parallel-stranded DNA quadruplex d(TTAGGGT)4 containing the human telomeric repeat. J. Mol. Biol. 334, 25–36.

Granzhan, A., Monchaud, D., Saettel, N., Guédin, A., Mergny, J.-L., Teulade-Fichou, M.-P., 2010. "One ring to bind them all"—Part II: identification of promising G-quadruplex ligands by screening of cyclophane-type macrocycles. J. Nucleic Acids, 460561.

Haider, S.M., Neidle, S., 2009. A molecular model for drug binding to tandem repeats of telomeric G-quadruplexes. Biochem. Soc. Trans. 37, 583–588.

Haider, S.M., Parkinson, G.N., Neidle, S., 2003. Structure of a G-quadruplex–ligand complex. J. Mol. Biol. 326, 117–125.

Hampel, S.M., Sidibe, A., Gunaratnam, M., Riou, J.-F., Neidle, S., 2010. Tetrasubstituted naphthalene diimide ligands with selectivity for telomeric G-quadruplexes and cancer cells. Bioorg. Med. Chem. Lett. 20, 6459–6463.

Harrison, R.J., Gowan, S.M., Kelland., L.R., Neidle, S., 1999. Human telomerase inhibition by substituted acridine derivatives. Bioorg. Med. Chem. Lett. 9, 2463–2468.

Hsu, C.F., Phillips, J.W., Trauger, J.W., Farkas, M.E., Belitsky, J.M., Heckel, A., et al., 2007. Completion of a programmable DNA-binding small molecule library. Tetrahedron 63, 6146–6151.

Irwin, J.J., Shoichet, B.K., 2005. ZINC—a free database of commercially available compounds for virtual screening. J. Chem. Inf. Modeling 45, 177–182.

Kim, M.Y., Vankayalapati, H., Shin-Ya, K., Wierzba, K., Hurley, L.H., 2002. Telomestatin, a potent telomerase inhibitor that interacts quite specifically with the human telomeric intramolecular G-quadruplex. J. Amer. Chem. Soc. 124, 2098–2099.

Lacroix, L., Séosse, A., Mergny, J.-L., 2010. Fluorescence-based duplex-quadruplex competition test to screen for telomerase RNA quadruplex ligands. Nucleic Acids Res. 2011 (39), e21.

Lee, H.M., Chan, D.S., Yang, F., Lam, H.Y., Yan, S.C., Che, C.M., et al., 2010. Identification of natural product fonsecin B as a stabilizing ligand of c-myc G-quadruplex DNA by high-throughput virtual screening. Chem. Commun. (Camb.) 46, 4680–4682.

Linder, J., Garner, T.P., Williams, H.E.L., Searle, M.S., Moody, C.J., 2011. Telomestain: formal total synthesis and cation-mediated interaction of its seco-derivatives with G-quadruplexes. J. Amer. Chem. Soc. 133, 1044–1051.

Lipinski, C.A., Lombardo, F., Dominy, B.W., Feeney, P.J., 2001. Experimental and computational approaches to estimate solubility and permeability in drug discovery and development settings. Adv. Drug Deliv. Rev. 46, 3–26.

Lombardo, C.M., Martínez, I.S., Haider, S., Gabelica, V., De Pauw, E., Moses, J.E., et al., 2010. Structure-based design of selective high-affinity telomeric quadruplex-binding ligands. Chem. Commun. (Camb.) 46, 9116–9118.

Ma, D.L., Lai, T.S., Chan, F.Y., Chung, W.H., Abagyan, R., Leung, Y.C., et al., 2008. Discovery of a drug-like G-quadruplex binding ligand by high-throughput docking. Chem. Med. Chem. 3, 881–884.

Martins, C., Gunaratnam, M., Stuart, J., Makwana, V., Greciano, O., Reszka, A.P., et al., 2007. Structure-based design of benzylamino-acridine compounds as G-quadruplex DNA telomere targeting agents. Bioorg. Med. Chem. Lett. 17, 2293–2298.

Mergny, J.-L., Lacroix, L., 2009. UV melting of G-quadruplexes. Curr. Protoc. Nucleic Acid Chem. Ch. 17, Unit 17.1.

Mergny, J.-L., Lacroix, L., Teulade-Fichou, M.-P., Hounsou, C., Guittat, L., Hoarau, M., et al., 2001. Telomerase inhibitors based on quadruplex ligands selected by a fluorescence assay. Proc. Natl. Acad. Sci. USA 98, 3062–3067.

Monchaud, D., Teulade-Fichou, M.-P., 2008. A hitchhiker's guide to G-quadruplex ligands. Org. Biomol. Chem. 6, 627–636.

Monchaud, D., Granzhan, A., Saettel, N., Guédin, A., Mergny, J.-L., Teulade-Fichou, M.-P., 2010. "One ring to bind them all"—Part I: the efficiency of the macrocyclic scaffold for G-quadruplex DNA recognition. J. Nucleic Acids, 525862.

Neidle, S., 2008. Principles of Nucleic Acid Structure. Academic Press, London.

Neilsen, M.C., Ulven, T., 2010. Macrocyclic G-quadruplex ligands. Curr. Med. Chem. 17, 3438–3448.

Paine, M.F., Wang, M.Z., Generaux, C.N., Boykin, D.W., Wilson, W.D., De Koning, H.P., et al., 2010. Diamidines for human African trypanosomiasis. Curr. Opin. Investig. Drugs 11, 876–883.

Paramasivan, S., Rujan, I., Bolton, P.H., 2007. Circular dichroism of quadruplex DNAs: applications to structure, cation effects and ligand binding. Methods 43, 324–331.

Parkinson, G.N., Cuenca, F., Neidle, S, 2008. Topology conservation and loop flexibility in quadruplex–drug recognition: crystal structures of inter- and intramolecular telomeric DNA quadruplex–drug complexes. J. Mol. Biol. 381, 1145–1156.

Parkinson, G.N., Ghosh, R., Neidle, S., 2007. Structural basis for binding of porphyrin to human telomeres. Biochemistry 46, 2390–2397.

Parkinson, G.N., Lee, M.P.H., Neidle, S., 2002. Crystal structure of parallel quadruplexes from human telomeric DNA. Nature 417, 876–880.

Pearce, H.L., Blanchard, K.L., Slapak, C.A., 2008. Failure modes in anticancer drug discovery and development. In: Neidle, S. (Ed.), Cancer Drug Design and Discovery. Academic Press, New York.

Pettersen, E.F., Goddard, T.D., Huang, C.C., Couch, G.S., Greenblatt, D.M., Meng, E.C., et al., 2004. UCSF Chimera—a visualization system for exploratory research and analysis. J. Comput. Chem. 25, 1605–1612.

Phan, A.T., Kuryavyi, V., Burge, S.E., Neidle, S., Patel, D.J., 2007. Structure of an unprecedented G-quadruplex scaffold in the human c-kit promoter. J. Amer. Chem. Soc. 129, 4386–4392.

Phan, A.T., Kuryavyi, V., Gaw, H.Y., Patel, D.J., 2005. Small-molecule interaction with a five-guanine-tract G-quadruplex structure from the human MYC promoter. Nature Chem. Biol 1, 167–173.

Rahman, K.M., Reszka, A.P., Gunaratnam, M., Haider, S.M., Howard, P.W., Fox, K.R., et al., 2009. Biaryl polyamides as a new class of DNA quadruplex-binding ligand. Chem. Commun. (Camb.), 4097–4099.

Ranjan, N., Andreasen, K.F., Kumar, S., Hyde-Volpe, D., Arya, D.P., 2010. Aminoglycoside binding to Oxytricha nova telomeric DNA. Biochemistry 49, 9891–9903.

Read, M., Harrison, R.J., Romagnoli, B., Tanious, F.A., Gowan, S.H., Reszka, A.P., et al., 2001. Structure-based design of selective and potent G-quadruplex-mediated telomerase inhibitors. Proc. Natl. Acad. Sci. USA 98, 4844–4849.

Riou, J.-F., Guittat, L., Mailliet, P., Laoui, A., Renou, E., Petitgenet, O., et al., 2002. Cell senescence and telomere shortening induced by a new series of specific G-quadruplex DNA ligands. Proc. Natl. Acad. Sci. USA 98, 2672–2677.

Sun, D., Thompson, B., Cathers, B.E., Salazar, M., Kerwin, S.M., Trent, J.O., et al., 1997. Inhibition of human telomerase by a G-quadruplex-interactive compound. J. Med. Chem. 40, 2113–2116.

Tan, J.H., Gu, L.Q., Wu, J.Y., 2008. Design of selective G-quadruplex ligands as potential anticancer agents. Mini Rev. Med. Chem. 8, 1163–1178.

Tisi, D., Chessari, G., Woodhead, A.J., Jhoti, H., 2008. Structural biology and anticancer drug design. In: Neidle, S. (Ed.), Cancer Drug Design and Discovery. Academic Press, New York.

Todd, A.K., Haider, S.M., Parkinson, G.N., Neidle, S., 2007. Sequence occurrence and structural uniqueness of a G-quadruplex in the human c-kit promoter. Nucleic Acids Res. 35, 5799–5808.

Wang, P., Leung, C.-H., Ma, D.-L., Yan, S.-C., Che, C.-M., 2010. Structure-based design of platinum (II) complexes as c-myc oncogene down-regulators and luminescent probes for G-quadruplex DNA. Chem. Eur. J. 16, 6900–6911.

Wang, Y., Patel, D.J., 1993. Solution structure of the human telomeric repeat d[AG$_3$(T$_2$AG$_3$)$_3$] G-tetraplex. Structure 1, 263–282.

Webba da Silva, M., 2007. NMR methods for studying quadruplex nucleic acids. Methods 43, 264–277.

10

The Determination of Quadruplex Structures

This chapter describes some of the background to the determination of the structures of quadruplex DNAs, RNAs and their small-molecule complexes that are discussed in this book—emphasizing diffraction and X-ray crystallography methodology, molecular visualization, structural databases, and key features of nucleotide conformation.

Fiber Diffraction Methods

Historically, helical DNA and RNA structures were first analyzed by fiber diffraction techniques, culminating in the double-helical model for DNA. Subsequent to 1953, this approach has been applied to both natural duplex nucleic acids and to a wide range of synthetic polynucleotides. The early work on guanosine-containing gels and simple repeating sequence such as poly (dG), as described in Chapters 1 and 2, was seminal in providing a firm conceptual base for the subsequent development of the subject. Both natural and synthetic polynucleotides can form fibers with varying degrees of internal order, having one- or two-dimensional para-crystalline arrays in the fiber, with the latter usually having the greater order because of their non-random sequences. These differing degrees of order are reflected in their X-ray diffraction patterns, with the purely guanine (or inosine)-containing polynucleotides in common with natural double-helical DNA and RNA molecules having a degree of order along the helix axis but being randomly oriented with respect to each other. This gives rise to an X-ray diffraction pattern with characteristic spots and streaks of intensity—for example, the "helical cross" diffraction pattern, which is characteristic of B-type DNA double helices. Such patterns can be analyzed to give the helical dimensions of pitch, rise and number of residues per helical turn, as well as defining the overall helical type. Even the best-ordered of para-crystalline polynucleotide fibers give at most only a few hundred individual diffraction maxima, corresponding to a typical maximum resolution of about 2.5 Å. It is not in general possible to analyze this pattern of fiber diffraction intensities, determine phases and derive a molecular structure *ab initio*, since the pattern is an average from all of the nucleotide units in a helical repeat. Instead, the pattern is fitted to a model using a least-squares procedure (Arnott, 1970). This enables conformational details of the averaged mono-repeat in poly (rG) or poly (dG) to be varied and optimized. The correctness and quality of the resulting model may be assessed using the standard crystallographic R factor (see below), which is a measure of agreement between observed structure amplitudes and

Therapeutic Applications of Quadruplex Nucleic Acids. DOI: 10.1016/B978-0-12-375138-6.00010-8

those calculated from the model. No fiber diffraction model for a four-stranded structure has been reported for some years, and the method is now of mainly historic interest. However, it should be possible to study fibers formed from short defined G-tract sequences and so the method has the potential to address questions of, for example, the arrangement formed by quasi-repeats of a sequence such as d[AGGG(TTAGGG)$_3$], if this can be induced to form a gel and then a fiber, analogous to the original analysis of 5′-GMP (see Chapters 1 and 2)

Single-crystal Methods

By contrast with fiber diffraction analyses, single-crystal X-ray crystallographic methods are able to determine the complete three-dimensional molecular structures of biological macromolecules without necessarily having recourse to any preconceived model, provided the molecules are discrete and not the effectively infinite polymeric helices of nucleic acid fibers. Crystallization of many DNA and RNA oligonucleotides has historically been challenging, and this appears to be even more the case with both DNA and RNA quadruplex sequences and their ligand complexes. Crystallization is often difficult and time-consuming, with a high rate of failures for a given sequence target. Whether these problems are associated with multiple species (conformational heterogeneity), or with poor crystal packing, is normally unanswerable. Instead the crystallographer has to rely upon systematic screening of sequence variants and mutants in order to arrive at tractable quadruplex-forming micro-crystal starting points. Some terminal sequences appear to be superior to others, although there are no definite rules as yet. A 5′-thymine or uracil sometimes aids crystallization, and a 5′-bromouracil derivative can further improve diffracting quality.

All approaches to quadruplex DNA and RNA crystallization, whether manual or automated, need to systematically screen a wide set of crystallizing conditions (Ducruix & Giegé, 1999; Campbell & Parkinson, 2007; Holbrook & Holbrook, 2010). It is now common practice to use automated robotic crystallization methods to rapidly set up and scan large numbers of crystallization trials, either in purely aqueous environments or under a thin layer of an oil. The robotic methods also have the advantage of using very small amounts of material, which is especially important when trying to crystallize RNA quadruplexes, given the greater expense and lower yields of RNA oligonucleotide synthesis. A number of pre-prepared crystallization kits have been developed with solutions containing a wide range of concentrations and types of counter-ion, buffer and precipitating agents, so that a very large number of crystallizing trials can be set up with minimal effort (for commercial sources of nucleic acid crystallization kits see, for example, http://hamptonresearch.com; http://www.mitegen.com). Specialized screening protocols have also been optimized for quadruplex crystallization (Campbell & Parkinson, 2007). The screening approach is important for finding initial crystallization conditions, and alternative crystal forms if initial trials produce crystals with poor diffraction, twinning or exceptionally large unit cells. In general crystallization experiments for quadruplex nucleic acids use any one of several "precipitants" to reduce the immediate water environment around a highly charged quadruplex. Most commonly this is a hydrophobic alcohol or a salt. Alcohols most often used are MPD (2-methyl-2,4-pentanediol) and, to a lesser extent, PEG (polyethylene alcohol). PEG is available in a wide range of molecular weights although a range of 400–4000 is normally employed. Ammonium sulfate has to date been reported only rarely as being a successful precipitant/crystallizing agent for quadruplexes. High cation concentrations are an essential component of the crystallization mix, which is unsurprising given the dependence of quadruplex stability on sodium or potassium ions, in particular.

It is perhaps coincidental that MPD, PEG and other agents have been used to induce a molecular crowding environment in a number of solution-based studies of quadruplex topology, which tends to produce parallel-type quadruplexes (see, for example, Xue *et al.*, 2007; Heddi & Phan, 2011). This is in agreement with the topology observed in crystal structures of native human telomeric quadruplexes (Parkinson, Lee & Neidle, 2002), rather than with the NMR structures in "dilute" solution. Although there are very little comparative data available as yet, it does appear that quadruplex–ligand complexes in solution in molecular crowding conditions also follow this pattern. An example is the porphyrin TMPyP4 (Martino *et al.*, 2009), whose telomeric quadruplex complex favors the more closely packed parallel topology, in accord with crystal-structure data (Parkinson, Ghosh & Neidle, 2007), albeit on a bimolecular TMPyP4–quadruplex complex. Thus it is tempting to conclude that crystal structures of human telomeric quadruplexes and their ligand complexes do reflect solution structures (and maybe even those in their biological environment), but that the solution condition needs also to be crowded or concentrated. It should not be assumed though that the use of MPD, PEG or other crowding agents, either in the conditions of crystal growth or in solution, in itself produces an innate tendency for parallel structures. The bimolecular quadruplex from *Oxytricha nova* has been crystallized as a native structure (Haider, Parkinson & Neidle, 2002) and with a number of ligands (for example, Campbell *et al.*, 2011), always in the presence of MPD. All these quadruplexes have an anti-parallel topology with diagonal loops, identical to that observed in solution by NMR methods in the absence of MPD (Schultze *et al.*, 1999).

High-intensity sources of X-rays are readily available at many high-intensity synchrotron sites worldwide, in addition to traditional laboratory X-ray facilities with rotating-anode and environmentally more acceptable and high-flux microfocus sealed-tube generators. Data collection at a synchrotron source is increasingly automated and often takes only minutes (or less) to collect a full data set, enabling large numbers of crystals to be screened. The range of resolution reported for single-crystal structures of DNA and RNA quadruplexes spans from 0.6 to >3.0 Å, with the majority being in the range 1.5–2.5 Å (Figure 10–1). The highest-resolution quadruplex structure, an RNA tetramolecular quadruplex (PDB id 1J8G) formed by four strands of r(UGGGGU) (Deng *et al.*, 2001), is at true atomic resolution, a rare occurrence, and is accordingly of corresponding high reliability and accuracy (≤0.02 Å for distances and ≤0.2° for angles) in respect of derived

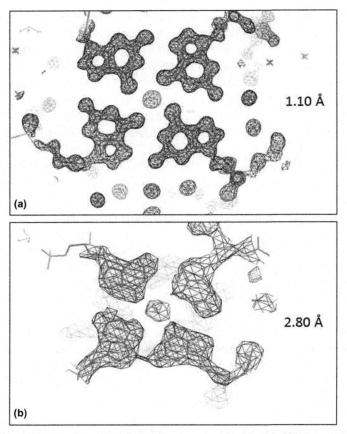

FIGURE 10–1 An electron density map in the plane of a G-quartet, calculated at differing resolutions, showing the amount of atomic detail visible at particular resolutions. **(a)** 1.10 Å, showing individual atoms at near-atomic resolution. **(b)** 2.80 Å, showing the overall shape of each guanine base, but not resolving individual substituent atoms. In each case the potassium ion at the center of the G-quartet can be clearly seen.

geometric parameters. A typical 2.5 Å resolution structure analysis, by contrast, would have distances reliable to about ±0.3 Å and angles to about ±5°. However, it is normally necessary to use constraints to standard bond geometries during the crystallographic refinement process of non-atomic resolution crystal structures. This means that it is not only non-bonded and intermolecular distances but also conformational and base morphological features that have to be interpreted with care, and likely errors and uncertainties taken into account. Hydrogen atoms are only directly observed in electron-density maps from the very highest-resolution quadruplex structure analyses, such as in PDB id 1J8G. So the donor–acceptor characteristics of hydrogen-bonding schemes (especially those involving water molecules) normally have to be inferred.

X-ray diffraction patterns from quadruplex oligonucleotide crystals can be in principle analyzed, and their underlying molecular structures solved *ab initio* by the standard heavy-atom multiple and single isomorphous replacement (MIR and SIR) phasing methods of macromolecular crystallography, although their use for quadruplex structures has not been reported, undoubtedly because of the power of the anomalous diffraction methods outlined below. All of these phasing methods share the advantage of not presuming any particular structural model and hence do not bias the resulting structure, for example, to have a particular quadruplex topology. A number of such heavy-atom derivatives are required for satisfactory MIR phasing, which are not always readily obtained, especially for quadruplex nucleic acids. In favorable cases it is possible to solve a structure with a single derivative by means of a combination of phasing from isomorphous replacement and anomalous scattering at a single wavelength, especially using the Single Anomalous Diffraction (SAD) method. The most common heavy-atom derivatization is the replacement of a methyl group in thymine with a bromine atom (resulting in 5-bromouracil). Bromine substituents can be sensitive to light and often decay in an intense X-ray beam, problems that are more acute with iodo derivatives. This can result in reduced occupancy of the attached nucleobase in the resulting electron-density maps.

The availability of tunable-wavelength X-ray facilities at many high-flux synchrotron facilities worldwide has enabled the technique of phasing by Multi-wavelength Anomalous Diffraction (MAD) to be used for the determination of quadruplex crystal structures. This uses the properties of a single appropriate heavy atom, which has the ability to absorb X-rays to differing extents at different wavelengths; phases and hence electron-density maps can be directly calculated from such data. These maps, when obtained at high resolution, are sometimes of remarkably high quality, revealing complete structures at the outset, with the important caveat that the inherent quality of the diffraction data, and hence of the crystal, must be very high. All too often oligonucleotide crystals (and quadruplexes are no exception) have high mosaicity and/or anisotropic diffraction, and exhibit diffuse scattering. It is fortunate for quadruplex nucleic acid crystallography that bromine atoms, which can be chemically attached to uracil bases, provide appropriate anomalous diffraction signals, so that several quadruplex crystal structures have been successfully solved by MAD methods, notably PDB ids 1J8G, 1K8P, 1V3N and 2GRB. In general a single bromine atom has sufficient anomalous phasing capability to phase structures containing some 25–30 nucleotides, so larger structures need to incorporate more than one bromine atom. This powerful approach is now the method of choice, limited only by the availability of suitable crystals rather than sufficient tunable synchrotron beam time. The optimal wavelength for the bromine anomalous edge is 0.92 Å, which is not normally available in in-house diffraction facilities. Several alternatives to the use of bromine have been found to be useful for other categories of nucleic acid, which are occasionally needed when it is found that the halogen–uracil bond is rapidly cleaved in the X-ray beam. Examples include using a single nucleotide with a thio-containing backbone (to bind a mercury heavy-atom derivative), or a phosphoroselenoate to replace a phosphate (Wilds *et al.*, 2002). It is possible to replace oxygen by selenium at the 2′ position of a thymine or uridine nucleotide (Jiang *et al.*, 2007; Pallan & Egli, 2007a), in the non-bridging backbone (Pallan & Egli, 2007b), or at the thymidine

4-position (Salon *et al.*, 2007). Oligonucleotides incorporating this selenium modification produce crystals that grow much more rapidly and have higher diffraction quality than do bromine-derivatized oligonucleotides.

It is in principle possible to utilize the anomalous signal of phosphorus atoms to phase nucleic acid structures when ultra-high-resolution (>1.0 Å) diffraction data of high redundancy are available (Dauter & Adamiak, 2001), although this has not as yet been applied to quadruplex DNAs or RNAs. More surprisingly, no report has as yet appeared of the use of a quadruplex channel metal ion for structure phasing, even though both strontium (Deng *et al.*, 2001) and thallium ions (Cáceres *et al.*, 2004; Gill, Strobel & Loria, 2006) have been located in the ion channels in two quadruplex crystal structures (PDB ids 2HBN and 1S45). Their potential has been demonstrated in the case of the d(TGGGGT)$_4$ structure (Cáceres *et al.*, 2004), where the Tl$^+$ ion was located by MAD methods, which enabled it to be distinguished from Na$^+$ ions. A similar method was used to locate and distinguish Ca^{2+} from Na$^+$ ions, also in a d(TGGGGT)$_4$ structure (Lee *et al.*, 2007).

Alternatively it is possible to take account of the fact that quadruplex structures can crystallize in an arrangement isomorphous to structures previously determined (for example, by heavy-atom phasing) or are presumed to contain a particular structural motif. These structures can often be solved by molecular replacement or "search" methods, which assume at least part of the structure and attempt to locate it in the crystallographic unit cell. Problems have occasionally arisen with this approach, when, for example, a molecule has been correctly oriented within the unit cell but its position is incorrectly indicated, being systematically related to the correct one, for example by a simple translation of a G-quartet. Search methods become increasingly challenging with a decreasing fraction of known geometry in a structure, and heavy-atom methods then become advisable. They are also difficult when the correct geometry of the search fragment is not precisely known, and then the correct rotational and translational solution becomes unclear. The use of NMR structures as starting points for molecular replacement seems obvious, but the success rate is low, probably because these methods are sensitive to even small (<1.0 Å) differences between a model and the correct crystal structure, which is well within the fluctuation range seen in an ensemble of NMR structures.

Macromolecular crystal structures are normally optimized with respect to the diffraction data by non-linear least-squares fitting procedures, which formally minimize the differences between observed and calculated models for the structure factors. This is the process of crystallographic refinement. When the diffraction data do not extend to atomic resolution, it is necessary to incorporate information from established stereochemical and structural features (such as bond lengths and angles, planar geometry of the DNA bases, preferred torsion angles). These are used to set up intramolecular constraints and restraints between them and so improve the initial models. Widely used programs for macromolecular refinement that are suitable for quadruplex structures include (i) X-PLOR/CNS (Brünger, Kreckowski & Erikson, 1990), which uses empirical energy terms as part of the minimized function to ensure optimal intra- and intermolecular geometry, (ii) REFMAC5 (Murshudov *et al.*, 1997), which minimizes structural coordinates and thermal parameters against a maximum likelihood residual. The technique of simulated annealing has been adopted from molecular dynamics as an effective way of refining structures when large-scale (>1 Å) atomic movements are required, since conventional least-squares methods are inherently incapable of effecting such large changes.

Quadruplex DNA and RNA crystal structures are heavily hydrated, with often over 50% solvent content. It is typical in medium-resolution structures for only a small fraction of these water molecules to be located in electron-density maps, largely because their high mobility smears their electron density to below the signal-to-noise level of these maps. The majority of water molecules reported in these structures are unsurprisingly the least mobile ones, which are directly hydrogen bonded to the structure—these are the "first-shell" water molecules (Figure 10–2). The ways in which molecules pack in the crystal are sometimes of importance when examining structural features, since considerations of efficient packing can readily force parts of molecules to interact one with another by hydrogen bonding and van der Waals interactions, and consequently possibly modify some features of otherwise flexible conformation.

The quality and reliability of a quadruplex DNA or RNA crystal structure are not straightforward to assess, especially for a non-crystallographer. Yet, judgments on these

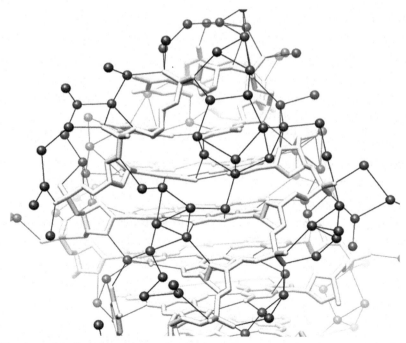

FIGURE 10–2 Structure of an *Oxytricha nova* quadruplex–ligand complex (Campbell *et al.*, 2011), at 1.10 Å resolution, viewed into one of the grooves and showing the water molecules (as small spheres) in the 1st and 2nd coordination shell. Hydrogen bonds are designated by thin lines.

factors are critical when undertaking and using structural comparisons and analyses. The important crystallographic parameters of quality (R, R_{free}) are defined as:

$$R = \sum \frac{||F_o| - |F_c||}{|F_o|}$$

summed over all observed reflections, where F_o and F_c are the observed and calculated structure factors.

R, which is also termed the reliability index, is expressed as a percentage, or sometimes as a decimal. The free R factor (R_{free}) is calculated in the same way but for a small (typically 5%) subset of reflections, often chosen randomly. This set is not used in the refinement, and so the value of R_{free} is unbiased by the course of the refinement and any errors introduced during it.

The R value for a correctly refined structural model can range between 10 and 20%. The lower the value the more reliable is the model. R_{free} values are often a few % higher. Since R_{free} is very sensitive to even small changes and errors in the model it is used as an indicator of the completion of a structure analysis, especially in terms of the behavior of water molecules located in electron-density maps and added to the model during successive rounds of refinement.

Of at least equal significance are the derived stereochemical features—examination of these is a reliable guide to quality (Das *et al.*, 2001).

Particular features to examine in a structure include:

- close non-bonded intra- and intermolecular contacts that are less than the sum of the van der Waals radii of the atoms involved;
- the distribution of values for torsion angles around single bonds. Eclipsed (~0°) values are indicative of problems in refinement;
- hydrogen bonds with distances significantly outside the accepted ranges of ~2.7–3.2 Å;
- estimates of error in atomic positions;
- the quality of the electron density for individual groups and atoms;
- values of atomic temperature factors, especially for water molecules.

In practice all new crystal structures are rigorously checked for consistency when they are being deposited in the PDB, and any problems are drawn to the attention of the investigators. However, it remains the case that a significant number of older structures retain some problematic features that have not been corrected.

Sources of Structural Data

The results of a crystal structure or fiber diffraction analysis are most useful as a set of atomic coordinates. Those from fiber diffraction are available either in the primary literature or in various review chapters and compilations; they are not deposited in the standard structural databases. Crystallographic coordinates may be obtained from a database, provided that they have been deposited in the first instance. This is no longer a problem since almost all journals now insist on deposition with publication, although authors in some journals still retain the right to up to a year's delay before public release of a data set. All available quadruplex crystal and NMR structures are in the RCSB (Research Collaboration for Structural Biology) Protein Data Bank (widely known as the PDB) (www.pdb.org), and at the European Bioinformatics Institute (EBI) (www.ebi.ac.uk/pdbe). All recent crystallographic depositions are accompanied by structure factor data. The Nucleic Acid Database (NDB) is a comprehensive relational database for nucleic acid crystallographic data, at Rutgers University, USA (Berman *et al.*, 1992) (http://ndbserver.rutgers.edu). It also provides a set of powerful tools for the comparative study of nucleic acid structural features, enabling detailed analysis of trends and features to be readily undertaken, although not all are fully functional for quadruplex nucleic acids.

Quadruplex Conformational Fundamentals

The five-membered deoxyribose and ribose sugar rings in DNA and RNA are inherently non-planar, i.e. are puckered. The precise conformation of a deoxyribose or ribose ring can be completely specified by the five endocyclic torsion angles within it (Figure 10–5). The ring puckering arises from the effect of non-bonded interactions between substituents at the four ring carbon atoms—the energetically most stable conformation for the ring has all substituents as far apart as possible. Thus different substituent atoms would be expected to produce differing types of puckering. The puckering can be described by either

- a simple qualitative description of the conformation in terms of atoms deviating from ring coplanarity; or
- precise descriptions in terms of the ring internal torsion angles.

In principle, there is a continuum of interconvertible puckers, separated by energy barriers. These various puckers are produced by systematic changes in the ring torsion angles. The puckers can be defined by the parameters P and τ_m, with the value of P, the phase angle of pseudorotation indicating the type of pucker since P is defined in terms of the five torsion angles τ_0–τ_4:

$$\tan P = \frac{(\tau_4 + \tau_1) - (\tau_3 + \tau_0)}{2*\tau_2*(\sin 36° + \sin 72°)}$$

and the maximum degree of pucker τ_m, by $\tau_m = \tau_2/\cos P$.

The pseudorotation phase angle can take any value between 0° and 360°. Values of τ_m indicate the degree of puckering of the ring. The five internal torsion angles are not independent of each other, and so to a good approximation any one angle τ_j can be represented in terms of just two variables: $\tau_j = \tau_m \cos [P + 0.8\pi(j - 2)]$.

Distinct deoxyribose ring pucker geometries have been observed experimentally, by X-ray crystallography and NMR techniques. When one ring atom is out of the plane of the other four, the pucker type is an envelope one. More commonly, two atoms deviate from the plane of the other three, with these two either side of the plane. It is usual for one of the two atoms to have a larger deviation from the plane than the other, resulting in a twist conformation. The direction of atomic displacement from the plane is important. If the major deviation is on the same side as the base and C4′–C5′ bond, then the atom involved is termed *endo*, and if on the opposite side, is termed *exo*. The most commonly observed puckers in crystal structures of quadruplexes are either close to C2′-*endo* or C3′-*endo* types (Figure 10–5). The C2′-*endo* family of puckers have P values in the range 140–185 and the C3′-*endo* domain has P values in the range −10 to +40°.

Sugar pucker preferences have their origin in the non-bonded interactions between substituents on the sugar ring and to some extent on their electronic characteristics. For example, the C3′-*endo* pucker would have hydroxyl substituents at the 2′ and 3′ positions further apart than with C2′-*endo* pucker; hence the preference of the former by RNA nucleic acids.

The bond between sugar and base in a nucleic acid is known as the glycosidic bond. Its stereochemistry is important. In natural nucleic acids the glycosidic bond is always β, that is the base is above the plane of the sugar when viewed onto the plane and therefore on the same face of the plane as the 5′ hydroxyl substituent. The absolute stereochemistry of other substituent groups on the deoxyribose (and ribose) sugar rings of DNA (and RNA) are defined such that when viewed end-on with the sugar ring oxygen atom O4′ at the rear (Figure 10–3), the hydroxyl group at the 3′ position is below the ring and the hydroxymethyl group at the 4′ position is above.

The glycosidic linkages are the C1′–N9 bond for purines and the C1′–N1 bond for pyrimidines. The torsion angle χ around this single bond can in principle adopt a wide range of values, although structural constraints result in marked preferences being observed. Glycosidic torsion angles are defined in terms of the four atoms:

O4′–C1′—N9—C4 for purines
O4′–C1′—N1—C2 for pyrimidines

Theory has predicted two principal low-energy domains for the glycosidic angle, in accord with experimental findings for a large number of nucleosides and nucleotides.

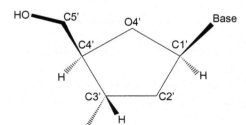

FIGURE 10–3 Stereochemistry of a deoxyribose sugar ring, showing the base and backbone attachments.

The *anti* conformation has the N1, C2 face of purines and the C2, N3 face of pyrimidines directed away from the sugar ring (Figure 10–4a) so that the hydrogen atoms attached to C8 of purines and C6 of pyrimidines are lying over the sugar ring. Thus, the Watson-Crick hydrogen-bonding groups of the bases are directed away from the sugar ring. These orientations are reversed for the *syn* conformation, with these hydrogen-bonding groups now oriented towards the sugar and especially its O5′ atom (Figure 10–4b). Crystal structures of *syn* purine nucleosides have found hydrogen bonding between the O5′ atom and the N3 base atom, which would stabilize this conformation. Otherwise, for purines, the *syn* conformation is slightly less preferred than the *anti*, on the basis of fewer non-bonded steric clashes in the latter case. The principal exceptions to this rule are guanosine-containing nucleotides, which have a small preference for the *syn* form because of favorable electrostatic interactions between the exocyclic N2 amino group of guanine and the 5′ phosphate atom. For pyrimidine nucleotides, the *anti* conformation is preferred over the *syn*, because of unfavorable contacts between the O2 oxygen atom of the base and the 5′-phosphate group.

The sterically preferred ranges for the two domains of glycosidic angles are:

anti: $-120 > \chi > 180°$
syn: $0 < \chi < 90°$

Values of χ in the region of about $-90°$ are often described as "high *anti*." There are pronounced correlations in quadruplex structures between sugar pucker and glycosidic angle (Neidle & Parkinson, 2008), which reflect the changes in non-bonded clashes produced by C2′-*endo* versus C3′-*endo* puckers. Thus, *syn* glycosidic angles are not found with C3′-*endo* puckers due to steric clashes between the base and the H3′ atom, which points towards the base in this pucker mode.

The phosphodiester backbone of an oligonucleotide has six variable torsion angles (Figure 10–6), designated $\alpha...\zeta$, in addition to the five internal sugar torsions $\tau_0...\tau_4$ and

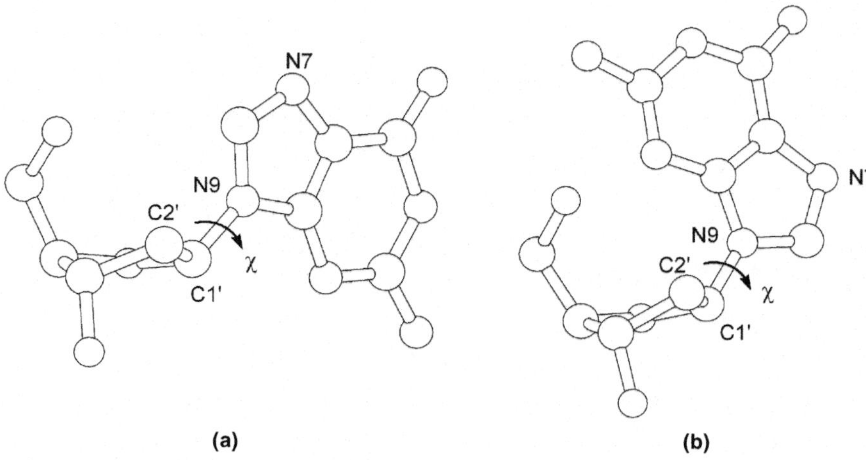

(a)   (b)

FIGURE 10–4 A guanosine deoxynucleoside in (a) *anti*, and (b) *syn* conformations.

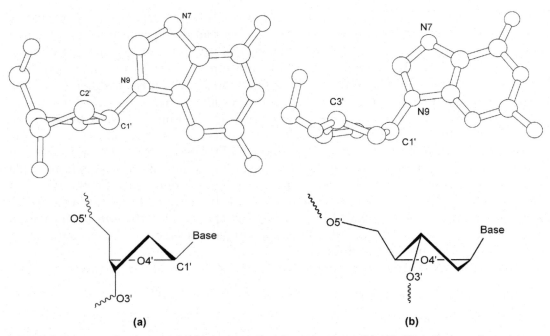

FIGURE 10–5 A guanosine deoxynucleoside in the *anti* conformation range, **(a)** with C2'-*endo* deoxyribose ring pucker, and **(b)** with C3'-*endo* pucker. Note the difference in glycosidic angle between the two puckers, resulting in distinct guanine base orientations.

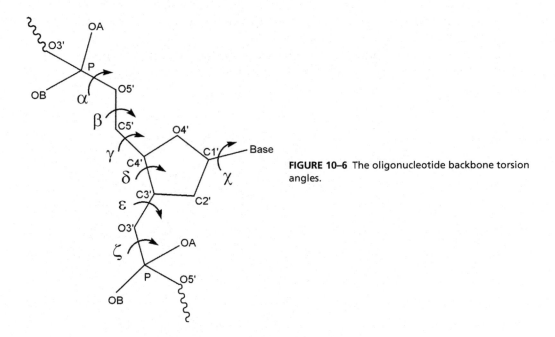

FIGURE 10–6 The oligonucleotide backbone torsion angles.

the glycosidic angle χ. A number of these have highly correlated values. Steric considerations alone dictate that the backbone angles are restricted to discrete ranges and are accordingly not free to adopt any value between 0 and 360°. A common convention for describing these backbone angles is to term values of ~60° as *gauche*⁺ (g⁺), −60° as *gauche*⁻ (g⁻) and ~180° as *trans* (t). Thus, for example, angles α (about the P–O5′ bond) and γ (the exocyclic angle about the C4′–C5′ bond) can be in the g⁺, g⁻ or t conformations.

There are a number of well-established correlations in duplex DNA involving pairs of these backbone torsion angles, as well as sugar pucker and glycosidic angle. Some of the more significant ones are relevant to quadruplexes, though exceptions are found (Neidle & Parkinson, 2008):

1. Between sugar pucker and glycosidic angle χ, especially for pyrimidine nucleosides. C3′-*endo* pucker is usually associated with median-value *anti* glycosidic angles, whereas C2′-*endo* puckers are commonly found with high *anti* χ angles. *Syn* glycosidic angle conformations show a marked preference for C2′-*endo* sugar puckers.
2. The C4′–C5′ torsion angle γ is correlated with the glycosidic angle and to some extent with sugar pucker and backbone angle α. *Anti* glycosidic angles tend to correlate with g⁺ conformations for angle γ.

Modeling of Quadruplex Structures

This is a rapidly expanding area, and the problems of adequately dealing with the dominant long-range electrostatic interactions in quadruplexes (Smirnov & Shafer, 2007) have now been largely overcome (Strahan, Keniry & Shafer, 1998; Šponer & Špačková, 2007). The simple molecular mechanics force field is still almost universally used, with the formalism being that the energetics of a molecule (V) is described in terms of the sum of a number of factors:

- 6–12 van der Waals non-bonded interactions;
- bond length and angle distortions;
- torsional barriers to rotation about single bonds;
- coulombic electrostatic contributions from full and partial electrostatic-potential derived atomic charges:

$$V(r^N) = \sum_{bonds} \frac{1}{2} k_b (l - l_0)^2 + \sum_{angles} \frac{1}{2} k_a (\theta - \theta_0)^2$$
$$+ \sum_{torsions} \frac{1}{2} V_n [1 + \cos(n\omega - \gamma)]$$
$$+ \sum_{j=1}^{N-1} \sum_{i=j+1}^{N} \left\{ 4\epsilon_{i,j} \left[\left(\frac{\sigma_{ij}}{r_{ij}} \right)^{12} - \left(\frac{\sigma_{ij}}{r_{ij}} \right)^6 \right] + \frac{q_i q_j}{4\pi\epsilon_0 r_{ij}} \right\}$$

The AMBER force field has been extensively parameterized for nucleic acids (see Pérez *et al.*, 2007 for a recent update of backbone parameters) and is the most commonly used one for quadruplex simulations, usually performed within the AMBER11 package (http://ambermd.org). Molecular dynamics simulations of a number of quadruplex systems have now been successfully performed and in large part they are able to reproduce experimental findings, not least the relative energetics of G-quartet base stacking in quadruplexes (see for a recent example, Cang, Šponer & Cheatham, 2011). Use of simulation methods for the modeling of quadruplex complexes involving new ligands is becoming an increasing popular approach to rational quadruplex ligand discovery; it is important to ensure that force-field parameters are not used that are inappropriate for nucleic acids.

Visualization of Quadruplex Molecular Structures

Molecular structures are best viewed interactively on a computer screen, rather than as flat representations on a page. This book provides Protein Data Bank identification codes, so that the reader can easily access and display individual structures, either by direct access to the PDB and the extensive visualization tools available from it, or by downloading coordinates from the PDB and inputting them into a molecular visualization or modeling program, such as one of those listed here.

A large number of such programs are now available, either commercially or as freeware. Many have been implemented on multiple platforms. Listed here are some recommended freeware programs that can be used on a PC (Windows™ or a Linux dialect) or a Macintosh™ desktop computer, laptop or iPad™ and which are nucleic acid structure-friendly in that they can cope with nucleic acid residues and can also display nucleic acid structures in cartoon form. Most of these programs can directly read files from the PDB on the web. Some molecular display software is also available as an app on the iPhone™.

1. **UCSF Chimera**, at www.cgl.ucsf.edu/chimera/. Developed by the Computer Graphics Laboratory, University of California, San Francisco, USA (Couch, Hendrix & Ferrin, 2006).
2. **VMD** (Visual Molecular Dynamics), at www.ks.uiuc.edu/Research/vmd/. Developed by the Theoretical and Molecular Biophysics Group, University of Illinois at Urbana, Champaign, USA (Humphrey, Dalke & Schulten, 1996).
3. **FirstGlance in Jmol**, an open source viewer, which can be run either as a stand-alone program or as a web browser. Obtainable from http://molvis.sdsc.edu/fgij or http://firstglance.jmol.org.
4. **Discovery Studio Visualizer** (PC and Linux only) is the freeware component of a comprehensive molecular modeling package. Obtainable from www.accelrys.com.

The Structures in this Book

Structures have been drawn using several different representations—sometimes more than one has been used in a single figure:

- Cartoons in which a phosphodiester backbone is represented by a ribbon (or by a tube), with the $5' \rightarrow 3'$ direction sometimes indicated by an arrow. Bases and sugars may be shown as filled slabs.
- Diagrams with bonds shown as lines or solid sticks.
- Figures with bonds shown as solid sticks and atoms as small spheres.
- van der Waals representations, having atoms drawn as spheres with radii set at their van der Waals values.
- Surface representations, showing the solvent-excluded surface of a molecule.

General Reading

Bloomfield, V.A., Crothers, D.M., Tinoco Jr., I., 2000. Nucleic Acids. Structures, Properties and Functions. University Science Books, Sausalito, California.

Blow, D.M., 2002. Outline of Crystallography for Biologists. Oxford University Press, Oxford.

Chayan, N.E. (Ed.), 2007. Protein Crystallization Strategies for Structural Genomics. International University Line

Egli, M., 2004. Nucleic acid crystallography: current progress. Curr. Opin. Struct. Biol. 8, 580–591.

Egli, M., Pallan, P.S., 2010. The many twists and turns of DNA: template, telomere, tool, and target. Curr. Opin. Struct. Biol. 20, 262–275.

Neidle, S. (Ed.), 1999. Oxford Handbook of Nucleic Acid Structure. Oxford University Press, Oxford.

Neidle, S., 2007. Principles of Nucleic Acid Structure. Academic Press, San Diego.

Rupp, B., 2009. Biomolecular Crystallography: Principles, Practice, and Application to Structural Biology. Garland Press, New York.

Saenger, W., 1984. Principles of Nucleic Acid Structure. Springer-Verlag, Berlin.

References

Arnott, S., 1970. The geometry of nucleic acids. Prog. Biophys. Mol. Biol. 21, 265–319.

Berman, H.M., Olson, W.K., Beveridge, D.L., Westbrook, J., Gelbin, A., Demeny, T., et al., 1992. The nucleic acid database. A comprehensive relational database of three-dimensional structures of nucleic acids. Biophys. J. 63, 751–759.

Brünger, A.T., Krukowski, A., Erikson, J., 1990. Slow-cooling protocols for crystallographic refinement by simulated annealing. Acta Crystallogr. A46, 585–593.

Cáceres, C., Wright, G., Gouyette, C., Parkinson, G., Subirana, J.A., 2004. A thymine tetrad in d(TGGGGT) quadruplexes stabilized with Tl^+/Na^+ ions. Nucleic Acids Res. 32, 1097–1102.

Campbell, N.H., Parkinson, G.N., 2007. Crystallographic studies of quadruplex nucleic acids. Methods 43, 252–263.

Campbell, N.H., Smith, D.L., Reszka, A.P., Neidle, S., O'Hagan, D., 2011. Fluorine in medicinal chemistry: β-fluorination of peripheral pyrrolidines attached to acridine ligands affects their interactions with G-quadruplex DNA. Org. Biomol. Chem. 9, 1328–1331.

Cang, X., Šponer, J., Cheatham III, T.E., 2011. Explaining the varied glycosidic conformational G-tract length and sequence preferences for anti-parallel quadruplexes. Nucleic Acids Res. 39, 4499–4512.

Couch, G.S., Hendrix, D.K., Ferrin, T.E., 2006. Nucleic acid visualization with UCSF Chimera. Nucleic Acids Res. 34, e29.

Das, U., Chen, S., Fuxreiter, M., Vaguine, A.A., Richelle, J., Berman, H.M., et al., 2001. Checking nucleic acid crystal structures. Acta Crystallogr. D57, 813–828.

Dauter, Z., Adamiak, D.A., 2001. Anomalous signal of phosphorus used for phasing DNA oligomer: importance of data redundancy. Acta Crystallogr. D57, 990–995.

Deng, J., Xioung, Y., Sundaralingam, M., 2001. X-ray analysis of an RNA tetraplex $(UGGGGU)_4$ with divalent $Sr^{(2+)}$ ions at subatomic resolution (0.61 Å). Proc. Natl. Acad. Sci. USA 98, 13665–13700.

Ducruix, A., Giegé, R. (Eds.), 2004. Crystallisation of Nucleic Acids and Proteins (second ed.). Oxford University Press, Oxford

Gill, M.L., Strobel, S.A., Loria, J.P., 2006. Crystallization and characterization of the thallium form of the *Oxytricha nova* G-quadruplex. Nucleic Acids Res. 34, 4506–4514.

Haider, S., Parkinson, G.N., Neidle, S., 2002. Crystal structure of the potassium form of an *Oxytricha nova* G-quadruplex. J. Mol. Biol. 320, 189–200.

Heddi, B., Phan, A.T., 2011. Structure of human telomeric DNA sequences in crowded conditions. J. Amer. Chem. Soc. DOI: 10.1021/ja2007869.

Holbrook, E.L., Holbrook, S.R., 2010. Crystallisation of nucleic acids. In: Encyclopedia of Life Sciences. John Wiley & Sons, Ltd, Chichester.

Humphrey, W., Dalke, A., Schulten, K., 1996. VMD—Visual Molecular Dynamics. J. Molec. Graphics 14, 33–38.

Jiang, J., Sheng, J., Carrasco, N., Huang, Z., 2007. Selenium derivatization of nucleic acids for crystallography. Nucleic Acids Res. 35, 477–485.

Lee, M.P., Parkinson, G.N., Hazel, P., Neidle, S., 2007. Observation of the coexistence of sodium and calcium ions in a DNA G-quadruplex ion channel. J. Amer. Chem. Soc. 129, 10106–10107.

Martino, L., Pagano, B., Fotticchia, I., Neidle, S., Giancola, C., 2009. Shedding light on the interaction between TMPyP4 and human telomeric quadruplexes. J. Phys. Chem. B 113, 14779–14786.

Murshudov, G.N., Vagin, A.A., Dodson, E.J., 1997. Refinement of macromolecular structures by the maximum-likelihood method. Acta Crystallogr. D53, 240–255.

Neidle, S., Parkinson, G.N., 2008. Quadruplex DNA crystal structures and drug design. Biochimie 90, 1184–1196.

Pallan, P.S., Egli, M., 2007. Selenium modification of nucleic acids: preparation of oligonucleotides with incorporated 2'-SeMe-uridine for crystallographic phasing of nucleic acid structures. Nature Protocols 2, 647–651.

Pallan, P.S., Egli, M., 2007. Selenium modification of nucleic acids: preparation of phosphoroselenoate derivatives for crystallographic phasing of nucleic acid structures. Nature Protocols 2, 640–646.

Parkinson, G.N., Ghosh, R., Neidle, S., 2007. Structural basis for binding of porphyrin to human telomeres. Biochemistry 46, 2390–2397.

Parkinson, G.N., Lee, M.P.H., Neidle, S., 2002. Crystal structure of parallel quadruplexes from human telomeric DNA. Nature 417, 876–880.

Pérez, A., Marchán, I., Svozil, D., Sponer, J., Cheatham 3rd, T.E., Laughton, C.A., et al., 2007. Refinement of the AMBER force field for nucleic acids: improving the description of alpha/gamma conformers. Biophys. J. 92, 3817–3829.

Salon, J., Sheng, J., Jiang, J., Chen, G., Caton-Williams, J., Huang, Z., 2007. Oxygen replacement with selenium at the thymidine 4-position for the Se base pairing and crystal structure studies. J. Amer. Chem. Soc. 129, 4862–4863.

Schultze, P., Hud, N.V., Smith, F.W., Feigon, J., 1999. The effect of sodium, potassium and ammonium ions on the conformation of the dimeric quadruplex formed by the *Oxytricha nova* telomere repeat oligonucleotide d($G(_4)T(_4)G(_4)$). Nucleic Acids Res. 27, 3018–3028.

Smirnov, I.V., Shafer, R.H., 2007. Electrostatics dominate quadruplex stability. Biopolymers 85, 91–101.

Šponer, J., Špa ková, N., 2007. Molecular dynamics simulations and their application to four-stranded DNA. Methods 43, 278–290.

Strahan, G.D., Keniry, M.A., Shafer, R.H., 1998. NMR structure refinement and dynamics of the K^+-[d($G_3T_4G_3$)]$_2$ quadruplex via particle mesh Ewald molecular dynamics simulations. Biophys. J. 75, 968–981.

Wilds, C.J., Pattanayek, R., Pan, C., Wawrzak, Z., Egli, M., 2002. Selenium-assisted nucleic acid crystallography: use of phosphoroselenoates for MAD phasing of a DNA structure. J. Amer. Chem. Soc. 124, 14910–14915.

Xue, Y., Kan, Z.Y., Wang, O., Yao, Y., Liu, J., Hao, Y.H., et al., 2007. Human telomeric DNA forms parallel-stranded intramolecular G-quadruplex in K+ solution under molecular crowding conditions. J. Amer. Chem. Soc. 129, 11185–11191.

Index

193

Printed in the United States
By Bookmasters